T0324692

ASPECTS OF
POLARON THEORY

Equilibrium and Nonequilibrium Problems

ASPECTS OF POLARON THEORY

Equilibrium and Nonequilibrium Problems

N. N. Bogolubov

N. N. Bogolubov Jr.

Moscow State University &
Russian Academy of Sciences, Russia

World Scientific

NEW JERSEY · LONDON · SINGAPORE · BEIJING · SHANGHAI · HONG KONG · TAIPEI · CHENNAI

Published by

World Scientific Publishing Co. Pte. Ltd.

5 Toh Tuck Link, Singapore 596224

USA office: 27 Warren Street, Suite 401-402, Hackensack, NJ 07601

UK office: 57 Shelton Street, Covent Garden, London WC2H 9HE

British Library Cataloguing-in-Publication Data
A catalogue record for this book is available from the British Library.

ASPECTS OF POLARON THEORY
Equilibrium and Nonequilibrium Problems

ISBN-13 978-981-283-398-3
ISBN-10 981-283-398-6

Printed in Singapore.

CONTENTS

Introduction

It is well known that a local change in the electronic state in a crystal leads to corresponding local changes in the interactions between individual atoms of the crystal, and hence to the excitation of atomic oscillations, i.e. the excitation of phonons. And vice versa, any local change in the state of the lattice ions alters the local electronic state. It is common in this situation to talk about an "electron–phonon interaction". This interaction manifests itself even at the absolute zero of temperature, and results in a number of specific microscopic and macroscopic phenomena. When an electron moves through the crystal, this state of polarization can move together with it. This combined quantum state, of "moving electron + accompanying polarization", may be considered as a sort of a quasiparticle with its own particular characteristics, such as effective mass, total momentum, energy, and maybe other quantum numbers describing the internal state of the quasiparticle in the presence of an external magnetic field presence or in the case of a very strong lattice polarization that causes self-localization of the electron in the polarization well with the appearance of discrete energy levels. Such a quasiparticle is usually called a "polaron state" or simply a "polaron". Polaron formation is a consequence of dynamic electron-lattice interaction which is also responsible for scattering of charge carriers, phonon frequency renormalization as well as screening of interaction between charge carriers in solids.

The concept of the polaron was introduced first by S.I. Pekar [1], who investigated the most essential properties of stationary polaron in the limiting case of very intense electron-phonon interaction, so that the polaron behavior could be analyzed in the so-called adiabatic approximation. Such famous researchers as L.D. Landau, S.I. Pekar, H. Fröhlich and R. Feynman have contributed to the development of polaron theory [1–5].

Despite the apparent simplicity of the formulation, the polaron problem has not yet been solved, and continues to attract much attention. It plays an important role in statistical mechanics and quantum field theory because it can be considered as the simplest example of a nonrelativistic quantum particle interacting with a quantum field. Therefore many sophisticated mathematical techniques have been tested for the first time using this problem as a model. A shining example of this is Feynman's functional integration method, which was applied first to the polaron problem, before becoming one of the main methods used in statistical mechanics and quantum field theory. Moreover, polaron theory is an expanding field of investigation in solid state physics because polarons are not only theoretical constructs but practically observable physical objects (see e.g. [6]).

One of the most important contributions to polaron theory, made by N.N. Bogolubov, is the rigorous adiabatic perturbation theory [7] created in 1950, in which the kinetic energy of the phonon field was treated as a small perturbation. The theory is translationally invariant (which is important for the development of the strong coupling theory), and reproduced at zeroth order the results for large values of the interaction constant that had already been derived. Despite a systematic attempt to calculate higher orders of the perturbation theory, these have not yet been derived, although much effort has been devoted to the problem.

Bogolubov returned to the polaron problem in 1980, when he developed and applied the well-known method of chronological or T-products [8]. This method appeared to be effective for the theory of the large-radius polarons for all strengths (weak, intermediate and strong) of electron–phonon interaction and also for the derivation of higher terms of the perturbation series in the weak-coupling limit. Like the functional integration formalism, the T-product method has various applications in many fields of quantum physics.

Interest to the polaron problem is growing: in addition to earlier fields of research dealing mostly with spatially homogeneous systems, investigation of charged-particle interactions with elementary excitations in spatially inhomogeneous low-dimensional systems, such as quantum wells, wires and boxes, is gaining significance. Experimental techniques have had great success in producing such systems with well-controlled parameters, thus allowing the manufacturing of structures with predictable characteristics. Electron–phonon interactions of the polaron type play a very important role in the properties of low-dimensional quantum systems. Thus, much efforts has been devoted to the investigation of surface polarons (see [9, 10] and references therein).

Of course, it is impossible to cover all off the numerous aspects of polaron theory in this short introduction or even in a far larger text. The main purpose of the present book is to acquaint the reader with methods of modern mathematical physics developed in connection with polaron theory.

The book is organized in the following way. Chapter 1 is an introduction to the T-product approach in the theory of a particle interacting with bosonic fields. As an example, this method is applied to the linearized polaron model and Feynman's two-body oscillator model, for which all calculations can be carried out explicitly. Feynman's well-known inequality in polaron theory is also reproduced as a particular case. The rest of the chapter is devoted to one version of finite-temperature perturbation theory for the polaron partition function and the ground-state energy developed on the basis of the T-product formalism. Adiabatic perturbation theory for the polaron ground-state energy, which is valid for the strong-coupling case, is also highlighted.

Chapter 2 deals with the equilibrium-state investigation for the Fröhlich polaron model. The main objective of this chapter is to derive Bogolubov's inequality for the reduced free energy of the polaron. This inequality allows one to obtain various upper bounds for the polaron ground-state energy relevant for different values of the particle–field interaction strength.

Chapter 3 touches on some problems related to nonequilibrium polaron theory including polaron kinetics. An exact evolution equation for a particle interacting with a bosonic field is derived here. It is shown that in the weak-coupling case this equation can be reduced to the Boltzmann equation in the polaron theory. Special attention is paid to the investigation of the nonequilibrium properties of the linearized polaron model. The main characteristics of this system, such as the impedance and the admittance, are calculated explicitly. It is also shown that the equilibrium momentum distribution function in the weak coupling limit can be derived by means of the T-product formalism without having recourse to the Boltzmann-equation approach.

Investigation of the dynamics in a "small" system weakly coupled to a "large" system (the heat bath) is one of the essential problems of statistical mechanics. The work by N. N. Bogolubov and N. M. Krylov [49] laid theoretical foundation for studies in this field. In this work the problem of possibility of a stochastic process in a dynamic system being under the influence of a large system was considered. The behavior of a classical system was studied on the basis of the Liouville equation for the probability distribution function in the phase space while for a quantum system the equivalent von Neuman equation for the statistical operator was employed. In [49] a method was developed allowing to derive the Fokker-Planck equation already in the first order approximation. In [50] a concrete model was studied in detail, the dynamics of which could be described by integrable equations. This property allowed rigorous critical analysis of various approximations to this model dynamics which had been derived earlier. Similar results for quantum mechanical systems were obtained in [51].

In lectures given by N. N. Bogolubov in 1974 while visiting the Rockefeller University, a modified version of the method, developed in [49], was outlined and its relation to the theory of two-time Green functions was discussed [52].

It is worth noticing that the notions of the "small" system and the "large" system are to be comprehended in the sense that the number of degrees of freedom of the former is much less than this number for the latter one.

Further development of ideas outlined in [49–52] provided an opportunity to formulate, on the basis of a model polaron problem, a method of derivation of exact system of hierarchic equations for time-dependent averages [35]. Bose-variables elimination from operator dynamic equations being averaged with respect to the initial statistical operator represents the cornerstone of the method. Special lemmas proved for the case of adiabatic switching on of the interaction between the "small" and the "large" systems played significant role in this elimination procedure [35]. If the "large" system is in the thermodynamically equilibrium state (being the heat bath in effect), the method allows to describe approach to equilibrium for the distribution of probabilities in the "small" system.

The above mentioned method proved itself very useful in studies of superradiant generation processes [53–55]. The phenomenon of superradiance reveals itself in appearance of spontaneous coherence of electromagnetic radiation due to photon exchange between atoms of active medium taking place under some additional conditions [56].

Here, in Chapter 3, an aproach to derive an exact equation for the evolution of a particle interacting with bosonic field is proposed. It is shown that in the case of weak interaction this equation can be reduced to the Boltzmann equation in the polaron theory. Particular attention is paid to investigation of nonequilibrium properties of the linearized polaron model. Principle characteristics of this model, such as impedance and admittance, are calculated explicitly. It is also shown that the equilibrium function of momentum distribution in the limiting case of weak interaction can be derived within the frame of the T-product formalism without any recourse to approximate Boltzmann equation.

Polaron Model: General Discussion

Let us consider a slow electron in a dielectric crystal, interacting with the lattice ions through long-range electrostatic forces. This electron will be permanently surrounded by a region of lattice polarization. Moving through the crystal, the electron carries the lattice distortion with it. The electron together with the accompanying self-consistent polarization field can be treated as a quasiparticle called a "polaron". Its effective mass is larger than that of a Bloch electron. Polaron formation is a consequence of the dynamical electron–phonon interaction.

One may speak about a "cloud of phonons" accompanying the electron. Thus a polaron can be also thought of as a compound system: "electron + accompanying phonons". The polaron problem was initially formulated in the context of solid state physics, where this concept has some direct applications [6, 11, 12]. On the other hand, this problem is of great theoretical interest quite apart from its particular solid-state interpretation, since it provides a very simple example of a particle interacting with a quantum field, and is thus a suitable model to probe the methods of quantum field theory and quantum statistics, and to formulate intuitive ideas about the properties of a particle moving through a fluctuating quantum medium. A detailed discussion on the physical origins and basic features of the polaron model can be found in old papers [11].

In this text we should like to give an introduction to a new method in the equilibrium polaron theory based on the T-product operator technique. Here and below we follow mainly the ideas outlined in our lectures [8].

Let us analyse in more detail the polaron Hamiltonian and its properties. From a general point of view, the polaron model may be considered as a particular case of a "small" subsystem S interacting with a "large" bosonic reservoir Σ. Let S be the electron and Σ be the phonon field of a crystal. Denote by X_S the set of arguments of the electron wave function and denote by $X_\Sigma = (...n_f...)$ the set of occupation numbers of the phonon modes.

The dynamical states of the polaron $S + \Sigma$ are then characterized by wave functions $\Psi(X_S, X_\Sigma)$ forming the space $\mathcal{H}_{S+\Sigma} = \mathcal{H}_S \otimes \mathcal{H}_\Sigma$, where $\mathcal{H}_S$ is the state space of the free electron while $\mathcal{H}_\Sigma$ is the phonon Fock space. We shall use below the notation $A(S)$, $A(\Sigma)$ and $A(S, \Sigma)$ for the operators acting correspondingly on the variables X_S, X_Σ and (X_S, X_Σ) of the wave function $\Psi(X_S, X_\Sigma)$. Note that the operators $A(S)$ and $A(\Sigma)$ will always commute with each other. The polaron Hamiltonian may be written as follows:

$$H_P = H(S) + H(\Sigma) + H_{\text{int}}(S, \Sigma), \tag{0.1}$$

with

$$H(S) = \frac{\mathbf{p}^2}{2m}, \tag{0.1a}$$

$$H(\Sigma) = \frac{1}{2} \sum_{(f)} (p_f p_{-f} + \omega_f^2 q_f q_{-f}), \tag{0.1b}$$

$$H_{\text{int}}(S, \Sigma) = \frac{1}{V^{1/2}} \sum_{(f)} L_f q_f e^{i f \cdot \mathbf{r}}, \tag{0.1c}$$

where the three operator terms correspond respectively to the Hamiltonian of the free band electron $H(S)$ with effective mass m, the Hamiltonian of the optical lattice phonons $H(\Sigma)$ with wave vectors $\mathbf{f}$ and frequencies ω_f, and the Hamiltonian of the electron–phonon interaction $H_{\text{int}}(S, \Sigma)$. The electron–phonon interaction is characterized by the coupling parameter L_f, which is assumed to be a real and spherically symmetric function:

$$L_f = L_f^* = L(|f|),$$

$\mathbf{r}, \mathbf{p}$ are quantum operators satisfying the usual commutation relations:

$$r_\alpha p_\beta - p_\beta r_\alpha = i\hbar \delta_{\alpha\beta} \qquad (\alpha, \beta = x, y, z).$$

he phonon amplitudes p_f, and q_f are also quantum operators satisfying analogous relations:

$$q_f p_{f'} - p_{f'} q_f = i\hbar \delta_{ff'},$$
$$p_f^\dagger = p_{-f}, \qquad q_f^\dagger = q_{-f}.$$

As usually, the phonon wave vector $\mathbf{f}$ runs over a quasidiscrete set of values:
$$\mathbf{f} = \left(\frac{2\pi n_1}{L}, \frac{2\pi n_2}{L}, \frac{2\pi n_3}{L}\right),$$

where $L^3 = V$ is the volume of the system and n_1, n_2, n_3 are integers.

In most articles the so-called Fröhlich polaron (also known as a large-radius polaron) is considered. For the Fröhlich polaron, the electron is supposed to interact with a dielectric continuum by means of long-range Coulomb forces. This assumption is adequate if the polaron, composed of the electron and the polarization well, which is induced by the electron itself, spreads over a range large compared with the lattice constant. Then, the polarization field $P(\mathbf{r})$ will be a smooth function of $\mathbf{r}$ and the polarization of the medium can be characterized by macroscopic dielectric constants ε_∞ and ε_0. The continuous approximation for the polarization field and hence the Fröhlich Hamiltonian itself would lose their meaning if the polaron size were comparable to the lattice constant.

The interaction parameter for the Fröhlich polaron model is determined in the following way:
$$L_f = \frac{g_0}{|\mathbf{f}|}, \qquad g_0 = e\left(\frac{1}{\varepsilon_\infty} - \frac{1}{\varepsilon}\right), \tag{0.2}$$

where e is the electron charge, ε_∞ and ε_0 are high-frequency and low-frequency dielectric constants. In the case of the usual Fröhlich model, one deals with the optical phonon branch, for which
$$\omega_f \to \omega > 0 \qquad \text{when} \qquad f \to 0,$$

and the dispersion is neglected, i.e. $\omega_f \equiv \omega$.

It is generally accepted that the strength of the interaction in this standard model can be characterized by a dimensionless coupling constant:
$$\alpha = \frac{g_0^2}{2\pi\hbar\omega^2}\left(\frac{m}{2\hbar\omega}\right)^{1/2}. \tag{0.3}$$

One usually distinguishes the cases of weak ($\alpha \ll 1$), strong ($\alpha > 10$) and intermediate ($\alpha \approx 3 - -6$) coupling.

It should be noted that when investigating the polaron problem in the general case, one should take into account the dependencies of ω_f and L_f on $\mathbf{f}$. In particular, one or other modification of the Coulomb case (0.2) might be analyzed. There are some physical reasons, for example, to introduce some kind of damping of the interaction for large $|\mathbf{f}|$. The simplest way to do this is to supplement the conditions (0.2) with the following restriction:
$$L_f \equiv 0, \tag{0.2a}$$

for $|\mathbf{f}| > f_0$, preserving old definition (0.2) for $|\mathbf{f}| < f_0$ at the same time. A natural value for the cut-off wave vector f_0 is $2\pi/a$, where a is the reciprocal lattice vector, since phonons with $|\mathbf{f}| > 2\pi/|\mathbf{a}|$ are not represented correctly in (0.1c) and thus can be omitted. Nevertheless, later we shall consider the standard Fröhlich polaron without any cut-off. And all calculations will be carried out, wherever possible, for an arbitrary functional dependence of the interaction parameter L_f.

Symmetries and Quantum Properties

It must be stressed that the polaron problem is essentially quantum in character. It may easily be shown, for instance, that for a "classical" electron the interaction (0.1c) is not important and reduces simply to some additive constant in the equivalent Hamiltonian.

Let us introduce in (0.1) instead of p_f and q_f the phonon creation and annihilation operators $b_f^\dagger$ and b_f:

$$q_f = \left(\frac{\hbar}{2\omega_f} \right)^{1/2} (b_f + b_{-f}^\dagger),$$

$$p_f = i \left(\frac{\hbar \omega_f}{2} \right)^{1/2} (b_f^\dagger - b_{-f}) \tag{0.4}$$

satisfying the commutation relations

$$b_f b_{f'}^\dagger - b_{f'}^\dagger b_f = \delta_{f,f'}, \quad b_f b_{f'} - b_{f'} b_f = 0, \quad b_f^\dagger b_{f'}^\dagger - b_{f'}^\dagger b_f^\dagger = 0. \tag{0.4a}$$

Then the Hamiltonian (0.1) reads

$$H_P = \frac{\mathbf{p}^2}{2m} + \sum_{(f)} \hbar \omega_f \left(b_f^\dagger b_f + \frac{1}{2} \right) + \frac{1}{V^{1/2}} \sum_{(f)} L_f \left(\frac{\hbar}{2\omega_f} \right)^{1/2} (b_f + b_{-f}^\dagger) e^{i f \cdot r}, \tag{0.5}$$

It may also be rewritten as

$$H_P = \frac{\mathbf{p}^2}{2m} + \sum_{(f)} \hbar \omega_f \left(B_f^\dagger B_f + \frac{1}{2} \right) - \frac{1}{V} \sum_{(f)} \frac{L_f^2}{2\omega_f^2}, \tag{0.6}$$

where

$$B_f = b_f e^{-i f \cdot r} + \frac{1}{V^{1/2}} \frac{L_f}{\hbar \omega_f} \left(\frac{\hbar}{2\omega_f} \right)^{1/2},$$

$$B_f^\dagger = b_f^\dagger e^{i f \cdot r} + \frac{1}{V^{1/2}} \frac{L_f}{\hbar \omega_f} \left(\frac{\hbar}{2\omega_f} \right)^{1/2}. \tag{0.6a}$$

The new operators B_f and $B_f^\dagger$ satisfy the same standard commutation relations as the Bose operators, (0.4a). If we assume that position and momentum operators $\mathbf{r}$ and $\mathbf{p}$ are commuting C-functions in the classical case then the operators B_f and $B_f^\dagger$ in (0.6) commute with the term $\mathbf{p}^2/2m$. In this case the relations (0.6a) can be interpreted as a canonical transformation to new Bose operators B_f and $B_f^\dagger$. Comparing (0.5) and (0.6), we conclude that for a classical electron the interaction is ineffective, being reduced to an additive constant term in the Hamiltonian.

On the contrary, in the quantum case, the "quasibosonic" amplitudes B_f and $B_f^\dagger$ do not commute with $\mathbf{p}^2/2m$ because of the factors $\exp(\pm i\mathbf{f}\cdot\mathbf{r})$, and hence the electronic and the quasibosonic parts of the Hamiltonian (0.6) are not independent of each other.

The difference between the classical and quantum situations may be clarified further by performing a unitary transformation on the Hamiltonian. To this end, let us introduce unitary operators

$$U = \exp\left(i\sum_{(f)} \mathbf{f}\cdot\mathbf{r}b_f^\dagger b_f\right), \qquad (0.7)$$

compensating for the exponential phase factors in B_f and $B_f^\dagger$:

$$UB_f U^\dagger = b_f + \frac{L_f}{\hbar\omega_f}\left(\frac{\hbar}{2\omega_f}\right)^{1/2} \equiv \tilde{b}_f, \qquad (0.8)$$

$$UB_f^\dagger U^\dagger = b_f^\dagger + \frac{L_f}{\hbar\omega_f}\left(\frac{\hbar}{2\omega_f}\right)^{1/2} \equiv \tilde{b}_f^\dagger. \qquad (0.9)$$

On the other hand the operator U transforms the electron momentum as follows:

$$U\mathbf{p}U^\dagger - \mathbf{p} - \sum_{(f)} \hbar\mathbf{f}h_f^\dagger h_f. \qquad (0.10)$$

The second term on the right-hand side here is obviously the total momentum of phonons. Thus we get, after U-transformation of the polaron Hamiltonian,[a]

$$H_P' = UH_P U^\dagger = \frac{1}{2m}\left(\mathbf{p} - \sum_{(f)}\hbar\mathbf{f}b_f^\dagger b_f\right)^2 + \sum_{(f)}\hbar\omega_f\left(\tilde{b}_f^\dagger\tilde{b}_f + \frac{1}{2}\right) - \sum_{(f)}\frac{|L_f|^2}{2\omega_f^2}$$

$$(0.10a)$$

$$= \frac{1}{2m}\left(\mathbf{p} - \sum_{(f)}\hbar\mathbf{f}b_f^\dagger b_f\right)^2 + \sum_{(f)}\hbar\omega_f b_f^\dagger b_f + \sum_{(f)}L_f\left(\frac{\hbar}{2\omega_f}\right)^{1/2}(b_f + b_{-f}^\dagger).$$

$$(0.10b)$$

Comparing (0.5) and (0.10b) and bearing in mind that the factors $\exp(\pm if\cdot\mathbf{r})$ in (0.5) are unimportant phase factors that are negligible in the classical case (in fact, they can be included in the operators $b_f^\dagger$ and b_f), we see that the quantum effect in polaron theory manifests itself in replacing the electron momentum $\mathbf{p}$ by the relative momentum of the electron with respect to the total momentum of phonons. It is interesting to note that the only important feature here is the quantum nature of the model itself but not the strength of the interaction.

---

[a] It is known that a unitary transformation of a Hamiltonian does not change the energy spectrum, and hence does not affect thermodynamic properties, ground-state energy, and so on.

Incidentally, it is possible to obtain some consequences of the representation (0.10) by observing that the momentum $\mathbf{p}$ is an integral of motion for the transformed Hamiltonian:

$$\mathbf{p}H'_P - H'_P\mathbf{p} \equiv 0.$$

On performing the inverse transformation, one finds that the corresponding integral of motion for the initial Hamiltonian is the sum of the electron momentum and the momentum of phonons:

$$\mathbf{P} = \mathbf{p} + \sum_{(f)} \hbar \mathbf{f} b_f^\dagger b_f, \qquad \mathbf{P}H_P - H_P\mathbf{P} \equiv 0. \qquad (0.11)$$

The latter identity can easily be verified by direct calculation. Note that the total momentum $\mathbf{P}$ is the generator of the translational symmetry group of the original Hamiltonian (0.5):

$$\mathbf{r} \to \mathbf{r} + \mathbf{a} \qquad (\mathbf{a} = \text{const}),$$

$$b_f \to b_f e^{-i\mathbf{a}\cdot\mathbf{f}}, \qquad b_f^\dagger \to b_f^\dagger e^{i\mathbf{a}\cdot\mathbf{f}}. \qquad (0.12)$$

The unitary transformations (0.7)–(0.12) were first introduced by Bogolubov [7] and Lee, Low and Pines [13] in order to develop appropriate approximate methods for the polaron problem.

Problems and Methods of Polaron Theory

One can distinguish two basic directions in polaron studies: the first deals with kinetic and transport properties, while the second investigates equilibrium properties, including quantum-mechanical phenomena at zero temperature.

In the kinetic theory one studies time-dependent phenomena in non-equilibrium or quasi-equilibrium situations such as relaxation processes (described by a Boltzmann-type equation) or the motion of an electron under given external forces, etc.

The equilibrium theory deals with the properties of the system at a given temperature. Of considerable interest are different averages related to the electron or to the polaron as a whole: the average kinetic and average total energies, the effective mass, the effective radius, etc. An interesting problem is to study the equilibrium distribution function of the electron momentum and its deviations from a Maxwellian form. Analogous problems can be formulated for the polaron ground state, which can also be considered as the limiting zero-temperature state (when $T \to 0$).

The basic function at equilibrium is the free energy (the logarithm of the partition function), which may be considered as a generating functional in order to compute the average energy and the ground-state energy. And, after introducing the corresponding fields (some additional terms) into the Hamiltonian, the free energy may be used to compute one

or another average, the effective mass, etc.[b] The equilibrium free energy
of a polaron will be the main quantity considered in the following sections.

Many important papers devoted to various aspects of the polaron
problem have been published. For the standard papers of the first period of
polaron studies one may consult [11] and references cited therein. Further
progress in the field is described in [6, 15, 37, 38, 59–61] and the numerous
references therein.

The general trend of developments can be seen from the titles of the
articles reproduced in the list of references, so we shall not review here all
the aspects of the polaron problem, making only some specific comments.
Examples of basic review articles are [6, 11, 17, 18].

Since the polaron Hamiltonian does not admit an exact solution, various approximate methods have been proposed in order to obtain numerical
results. These methods usually involve elements of perturbation theory,
canonical transformations and variational principles.

At zero temperature the polaron problem is a quantum-mechanical
problem (see [11] and references therein). In the weak-coupling case $\alpha \ll 1$
one can apply a more or less standard perturbation approach. Some
improvements of the perturbation scheme can be achieved by appropriate
canonical transformation of the Hamiltonian and a proper choice of the
trial (variational) wave function (see [7, 11, 13, 19–24]). Special forms of
perturbation theory have also been developed for the strong-coupling case
[2, 7, 11, 20, 25–27].

The problem becomes more complicated when one investigates polaron
equilibrium properties at finite temperature [5, 28].

Analogous investigations have also been performed for the nonequilibrium situation in [29–32].

A new general method in the polaron theory has been proposed by
N.N. Bogolubov and N.N. Bogolubov, Jr. in [8, 33–35], which is based
on the elimination of the phonon degrees of freedom by means of the

---

[b] For an arbitrary system at equilibrium, with the temperature
$\vartheta = kT = \beta^{-1}$ and Hamiltonian H, the free energy is given by

$$f[H, \beta] = -\frac{1}{\beta} \ln \operatorname{Tr} e^{-\beta H},$$

where $\beta = \vartheta^{-1}$ is the inverse temperature. Here $\operatorname{Tr} e^{-\beta H}$ is the so-called partition function. An arbitrary average $\langle A \rangle_{\beta,H}$ can be obtained, in particular, by
differentiating the free energy with respect to the corresponding source term
introduced into the Hamiltonian:

$$\langle A \rangle_{\beta,H} = -\frac{\partial}{\partial x} f[\beta, H - xA]_{x=0}.$$

For instance, the average energy is

$$\langle H \rangle_{\beta,H} = \frac{\partial}{\partial \beta} \{\beta f[H, \beta]\}.$$

averaged T-product operator technique. This technique may be considered as an analog of the path-integration approach. However, it seems to be much more transparent and rigorous from the mathematical point of view, and more convenient for practical calculations. For instance, when treating equilibrium aspects of a polaron below, we shall deal in all cases with proper quantum Gibbs averages over quadratic bosonic Hamiltonians instead of cumbersome path integral analysis.

A generalized approximation scheme for the free energy (partition function) has been developed in [8], based on a "linear-model" trial Hamiltonian. The linear model can be considered as a natural generalization of Feynman's two-body approximation (0.13) to the case of a continuum of "heavy particles" coupled with the electron. All of the characteristics of the linear-model Hamiltonian can be evaluated exactly in terms of the spectral representation, thus providing the basis for a systematic variational approach in a general form. In [36] a perturbation theory for the free energy (partition function) has been considered within the framework of the T-product approach. The T-product approach has also been developed for the nonequilibrium case. In [33, 34] a generalized kinetic equation with eliminated phonons has been derived. After simple approximations this equation yields the standard Boltzmann equation for a polaron, and may be used to obtain its generalizations. Some other applications of the generalized kinetic equation with eliminated phonons can be found in [35]. We should also mention that in [35]. the expression for the impedance and the admittance of a polaron, derived earlier in [29] by path integration, are reproduced on the basis of the linear-model Hamiltonian in a simpler and more rigorous manner.

In [59] a linear polaron model in constant uniform magnetic field was considered at zero temperature. An approach based on the model Hamiltonian diagonalization by means of the Bogolubov u–v transformation was proposed. The ground state energy was studied in the simplest case of equal frequencies for all the phonons involved in the interaction. Joint effect of the magnetic field and the electron-phonon interaction on the energy spectrum was studied too. It was also shown that the usage of the linear model as a trial model results in the action functional commonly employed in treatment of polarons in external constant uniform magnetic field [57, 58].

Chapter 1

LINEAR POLARON MODEL

This chapter is mainly a pedagogical introduction to a modern method in equilibrium finite-temperature polaron theory based on the T-product operator technique. As an example of the application of this method, Bogolubov's exactly solvable linearized polaron model, as well as Feynman's two-body oscillator model, are considered.

1.1. Introduction to the Linear Polaron Model

Here we consider the so-called linear polaron model described by a Hamiltonian that consists of the well-known oscillator Hamiltonian H_S, the phonon field Hamiltonian H_Σ and the interaction Hamiltonian $H_{S\Sigma}$, i.e.

$$H_{S+\Sigma} = H_S + H_\Sigma + H_{S\Sigma}, \qquad (1.1)$$

where

$$H_S = \frac{\mathbf{p}^2}{2m} + \frac{K^2 \mathbf{r}^2}{2}, \qquad H_\Sigma = \frac{1}{2} \sum_{(f)} \{p_f p_f^* + \nu^2(f) q_f q_f^*\},$$

$$H_{S\Sigma} = \frac{1}{V^{1/2}} \sum_{(f)} S(f) \mathbf{f} \cdot \mathbf{r} q_f.$$

Here $\mathbf{r}$ and $\mathbf{p}$ are respectively the position and the momentum of the electron and $S(f) = S(|f|)$ is a real radially symmetric function:

$$\nu(f) = \nu(|f|) > 0,$$

$$q_{-f} = q_f^*, \qquad p_{-f} = p_f^*.$$

Summation over f is over the range of quasidiscrete values

$$f = \left(\frac{2\pi n_1}{L}, \frac{2\pi n_2}{L}, \frac{2\pi n_3}{L} \right),$$

where $L^3 = V$ is the volume of the system and n_1, n_2, n_3 are integers covering the whole space of integers from $-\infty$ to $+\infty$. The total number of oscillators N is assumed to be finite for any finite volume V (later we should take the so-called thermodynamic limit as usual; that is, we must put $N \to \infty$, $V \to \infty$, imposing the additional condition $N/V = $ const).

It should be noted that the following identity holds:

$$\frac{1}{2}\sum_{(f)}\nu^2(f)\left\{q_f - \frac{i\mathbf{f}\cdot\mathbf{r}}{\nu^2(f)}\frac{S(f)}{V^{1/2}}\right\}\left\{q_f^* + \frac{i\mathbf{f}\cdot\mathbf{r}}{\nu^2(f)}\frac{S(f)}{V^{1/2}}\right\} = \frac{1}{2}\sum_{(f)}\nu^2(f)q_f q_f^*$$

$$+ \frac{1}{2}\sum_{(f)}\frac{1}{\nu^2(f)}\frac{S^2(f)}{V}(\mathbf{f}\cdot\mathbf{r})^2 + \frac{i}{2}\sum_{(f)}(\mathbf{f}\cdot\mathbf{r}q_f - \mathbf{f}\cdot\mathbf{r}q_f^*)\frac{S(f)}{V^{1/2}}.$$

It is obvious that

$$-i\sum_{(f)}\mathbf{f}\cdot\mathbf{r}q_f^*\frac{S(f)}{\sqrt{V}} = -\frac{i}{V^{1/2}}\sum_{(f)}\mathbf{f}\cdot\mathbf{r}q_{-f}S(-f) = \frac{1}{V^{1/2}}\sum_{(f)}\mathbf{f}\cdot\mathbf{r}q_f S(f),$$

where the property $q_{-f} = q_f^*$ has been taken into account.

Because of the radial symmetry, the following identity holds for an arbitrary function $F(|f|)$:

$$\sum_{(f)}F(|f|)f_\alpha f_\beta = \delta_{\alpha\beta}\frac{1}{3}\sum_{(f)}F(|f|)f^2,$$

where

$$|f| = (f_1^2 + f_2^2 + f_3^2)^{1/2}, \qquad f^2 = |f|^2.$$

Therefore

$$\frac{1}{2V}\sum_{(f)}\frac{S^2(f)}{\nu^2(f)}(\mathbf{f}\cdot\mathbf{r})^2 = \frac{r^2}{6V}\sum_{(f)}\frac{S^2(f)f^2}{\nu^2(f)}.$$

Therefore the potential energy can be represented in the form

$$U = \frac{K^2 r^2}{2} + \frac{1}{2}\sum_{(f)}\nu^2(f)q_f q_f^* + \frac{i}{V^{1/2}}\sum_{(f)}S(f)\mathbf{f}\cdot\mathbf{r}q_f = \frac{(K^2 - K_0^2)r^2}{2}$$

$$+ \frac{1}{2}\sum_{(f)}\nu^2(f)\left\{q_f - \frac{i\mathbf{f}\cdot\mathbf{r}}{\nu^2(f)}\frac{S(f)}{V^{1/2}}\right\}\left\{q_f^* + \frac{i\mathbf{f}\cdot\mathbf{r}}{\nu^2(f)}\frac{S(f)}{V^{1/2}}\right\}, \quad (1.2)$$

where

$$K_0^2 = \frac{1}{3V}\sum_{(f)}\frac{S^2(f)f^2}{\nu^2(f)}.$$

Consider the case $K^2 = K_0^2$. For this case, $U \geqslant 0$, with $U = 0$ if

$$q_f = \frac{i\mathbf{f}\cdot\mathbf{r}}{\nu^2(f)}\frac{S(f)}{V^{1/2}}.$$

We write down the kinetic energy which is obviously positive:

$$\frac{\mathbf{p}^2}{2m} + \frac{1}{2}\sum_{(f)} |p_f|^2.$$

Let us introduce normal variables Q_λ. Then $\mathbf{r}$ and q_f are linear combinations of the new variables Q_λ. It will be noted that for $K = K_0$ the system (1.1) is described by the Hamiltonian $H = T + U$, where both quadratic forms are positive-definite. It can be shown by purely linear algebraic methods that these can be reduced to the diagonal form simultaneously, and the Hamiltonian reads as

$$H = \sum_\lambda \left(\dot{Q}_\lambda^2 + \Omega_\lambda^2 Q_\lambda^2 \right).$$

if written in the new normal variables Q_λ. In this case each Q_λ satisfies the following equation:

$$\ddot{Q}_\lambda + \Omega_\lambda^2 Q_\lambda = 0.$$

For $K = K_0$, it follows from (1.2) that U becomes zero if and only if all q_f belong to the three-dimensional set

$$q_f = \frac{i\mathbf{f} \cdot \mathbf{r}}{\nu^2(f)} \frac{S(f)}{V^{1/2}}.$$

In other words, for $K = K_0$ the Hamiltonian H is translation-invariant with respect to the three-dimensional group of translations:

$$\mathbf{r} \to \mathbf{r} + \mathbf{R}, \qquad q_f \to q_f + \frac{i\mathbf{f} \cdot \mathbf{R}}{\nu^2(f)} \frac{S(f)}{V^{1/2}}.$$

Hence exactly three components among the whole set of Ω_λ^2 are equal to zero, while the other Ω_λ^2 are positive. So, there are three modes of collective evolution, such that

$$\ddot{Q}_\alpha = 0,$$

which correspond to inertial motion. Therefore $\mathbf{r}(t)$ describes uniform inertial motion in the case $K = K_0$, on which harmonic vibrations are superimposed.

Note also that when $K < K_0$, the form U is not positive, so that some values Ω_λ^2 must be negative, and the motion is unstable and can be characterized by the exponentially increasing function of t.

Later we will be interested especially in the case $K = K_0$, but it is more convenient for technical reasons to consider the more general expression

$$K^2 = K_0^2 + \eta^2, \tag{1.3}$$

having in mind a future passage to the limit $\eta \to 0$ (which must be taken before the usual limit $V \to \infty$).

It is worth stressing that, with the above choice for K, the form U is positive-definite because $\eta^2 > 0$ by definition. So, all values Ω_λ^2 are positive too. Therefore all of the functions $\mathbf{r}(t)$, $\mathbf{p}(t)$, $q_f(t)$ and $p_f(t)$ can be represented as corresponding sums of harmonic vibrations.

1.2. Equations of Motion

Let us introduce Bose amplitudes b_f and $b_f^\dagger$ by means of the relations

$$q_f = \left(\frac{\hbar}{2\nu(f)}\right)^{1/2} (b_f + b_{-f}^\dagger), \qquad p_f = i\left(\frac{\hbar\nu(f)}{2}\right)^{1/2} (b_f^\dagger - b_{-f}).$$

These amplitudes satisfy the usual commutation relations

$$b_f b_f^\dagger - b_f^\dagger b_f = 1.$$

One can see from here that

$$q_{-f} = q_f^*, \qquad p_{-f} = p_f^*,$$

and also

$$[q_f, p_f] = \frac{q_f p_f - p_f q_f}{i\hbar} = 1.$$

Therefore the Hamiltonian (1.1) can be rewritten in the form

$$H = \frac{\mathbf{p}^2}{2m} + \frac{1}{2}(K_0^2 + \eta^2)\mathbf{r}^2 + i\sum_{(f)} \frac{1}{V^{1/2}} S(f) \left(\frac{\hbar}{2\nu(f)}\right)^{1/2} \mathbf{f} \cdot \mathbf{r}(b_f + b_{-f}^\dagger)$$

$$+ \sum_{(f)} \hbar\nu(f) b_f^\dagger b_f + \frac{1}{2} \sum_{(f)} \hbar\nu(f). \quad (1.4)$$

The equations of motion for this Hamiltonian are

$$\frac{d\mathbf{r}}{dt} = \frac{\partial H}{\partial \mathbf{p}}, \qquad \frac{d\mathbf{p}}{dt} = -\frac{\partial H}{\partial \mathbf{r}},$$

$$i\hbar\frac{db_f}{dt} = b_f H - H b_f, \qquad i\hbar\frac{db_{-f}^\dagger}{dt} = b_{-f}^\dagger H - H b_{-f}^\dagger.$$

Transforming the right-hand sides of these equations, we see that

$$m\frac{d\mathbf{r}}{dt} = \mathbf{p},$$

$$\frac{d\mathbf{p}}{dt} = -(K_0^2 + \eta^2)\mathbf{r} - \frac{i}{V^{1/2}} \sum_{(f)} S(f) \left(\frac{\hbar}{2\nu(f)}\right)^{1/2} \mathbf{f}(b_f + b_{-f}^{\dagger}),$$

$$i\hbar\frac{db_f}{dt} = \hbar\nu(f)b_f - \frac{i}{V^{1/2}} \left(\frac{\hbar}{2\nu(f)}\right)^{1/2} S(f)\mathbf{f}\cdot\mathbf{r},$$

(1.5)

$$i\hbar\frac{db_{-f}^{\dagger}}{dt} = -\hbar\nu(f)b_{-f}^{\dagger} + \frac{i}{V^{1/2}} \left(\frac{\hbar}{2\nu(f)}\right)^{1/2} S(f)\mathbf{f}\cdot\mathbf{r}.$$

Later we are going to convert the system of equations (1.5) into a system of equations for Green functions. This step allows us to calculate explicitly such Green functions as $\langle\langle r_\alpha, r_\beta \rangle\rangle$ and $\langle\langle p_\alpha, p_\beta \rangle\rangle$, the spectral function $J_{p_\alpha p_\beta}$, and hence the equilibrium correlation functions $\langle p_\alpha(t)p_\beta(\tau)\rangle$ calculated with respect to the Hamiltonian (1.4).

It is appropriate to recall the definition of the two-time correlation and Green functions [40, 41]. For any two operators $A(t)$ and $B(\tau)$ taken in the Heisenberg representation, two-time equilibrium correlation functions are usually defined in the following way:

$$\langle A(t)B(\tau)\rangle_{\text{eq}} = \int\limits_{-\infty}^{+\infty} J_{A,B}(\omega)e^{-i\omega(t-\tau)}\, d\omega,$$

(1.6)

$$\langle B(\tau)A(t)\rangle_{\text{eq}} = \int\limits_{-\infty}^{+\infty} J_{A,B}(\omega)e^{-\beta\omega\hbar}e^{-i\omega(t-\tau)}\, d\omega,$$

where

$$\beta = \frac{1}{\vartheta} = \frac{1}{K_\beta T},$$

and K_β is Boltzmann's constant and T the absolute temperature. Retarded and advanced Green functions can be introduced in the usual manner [41]:

$$\langle\langle A(t), B(\tau)\rangle\rangle_{\text{ret}} = \vartheta(t-\tau)\langle[A(t), B(\tau)]\rangle_{\text{eq}}$$

$$= \vartheta(t-\tau)\frac{\langle A(t)B(\tau) - B(\tau)A(t)\rangle_{\text{eq}}}{i\hbar}, \quad (1.7)$$

$$\langle\langle A(t)B(\tau)\rangle\rangle_{\text{adv}} = -\vartheta(\tau-t)\langle[A(t), B(\tau)]\rangle_{\text{eq}}.$$

Here $\langle\ldots\rangle_{\text{eq}}$ denotes the statistical-equilibrium average value calculated with respect to the Hamiltonian (1.4):

$$\langle\ldots\rangle_{\text{eq}} = \frac{\text{Tr } e^{-H/\vartheta}(\ldots)}{\text{Tr } e^{-H/\vartheta}}.$$

Introduce a function of the complex variable Ω, $\text{Im }\Omega \neq 0$:

$$\langle\langle A, B\rangle\rangle_\Omega = \frac{1}{\hbar}\int\limits_{-\infty}^{+\infty} J_{A,B}(\nu)\frac{1 - e^{-\beta\nu\hbar}}{\Omega - \nu}\, d\nu.$$

(1.8)

Then the spectral densities for the advanced and retarded Green functions can be introduced by means of the relations

$$\langle\langle A(t), B(\tau)\rangle\rangle_{\text{adv}} = \frac{1}{2\pi} \int\limits_{-\infty}^{+\infty} \langle\langle A, B\rangle\rangle_{\omega - i0} e^{-i\omega(t-\tau)} \, d\omega,$$

$$\langle\langle A(t), B(\tau)\rangle\rangle_{\text{ret}} = \frac{1}{2\pi} \int\limits_{-\infty}^{+\infty} \langle\langle A, B\rangle\rangle_{\omega + i0} e^{-i\omega(t-\tau)} \, d\omega. \tag{1.9}$$

Taking into account the well-known formula

$$\frac{1}{\omega - \nu \pm i\varepsilon} = \mathcal{P}\left(\frac{1}{\omega - \nu}\right) \mp i\pi\delta(\omega - \nu), \tag{1.10}$$

we arrive at the important relation

$$\langle\langle A, B\rangle\rangle_{\omega + i0} - \langle\langle A, B\rangle\rangle_{\omega - i0} = -\frac{2\pi i}{\hbar} J_{A,B}(\omega)(1 - e^{-\beta\hbar\omega}). \tag{1.11}$$

Our aim is to derive a system of equations for the Green functions (1.7). From a formal point of view,

$$\frac{d}{dt}\vartheta(t) = \delta(t), \qquad \frac{d}{dt}\vartheta(-t) = -\delta(t).$$

These relations allows us to differentiate formally both sides of (1.7), thus leading to the desired equations for the Green functions:

$$i\frac{d}{dt}\langle\langle A(t), B(\tau)\rangle\rangle_{\text{ret, adv}}$$

$$= \frac{1}{\hbar}\delta(t-\tau)\langle AB - BA\rangle + \left\langle\left\langle i\frac{dA(t)}{dt}, B(\tau)\right\rangle\right\rangle_{\text{ret, adv}}. \tag{1.12}$$

From these equations one has, in the "Ω-representation"

$$\Omega\langle\langle A, B\rangle\rangle_{\Omega} = \frac{1}{\hbar}\langle AB - BA\rangle + \left\langle\left\langle i\frac{dA}{dt}, B\right\rangle\right\rangle_{\Omega}, \tag{1.13}$$

or equivalently, in a slightly different form,

$$-i\Omega\langle\langle A, B\rangle\rangle_{\Omega} = \frac{1}{i\hbar}\langle AB - BA\rangle + \left\langle\left\langle \frac{dA}{dt}, B\right\rangle\right\rangle_{\Omega}.$$

We should note one more useful identity[c]:

$$\langle B(\tau)A(t)\rangle_{\text{eq}} = \int\limits_{-\infty}^{+\infty} J_{B,A}(\omega)e^{-i\omega(\tau-t)}\,d\omega = \int\limits_{-\infty}^{+\infty} J_{B,A}(-\omega)e^{-i\omega(t-\tau)}\,d\omega,$$

where

$$J_{B,A}(-\omega) = J_{A,B}(\omega)e^{-\beta\omega\hbar} \qquad \text{(cf. (1.6)),}$$

From (1.8), one has, after the permutation

$$\begin{pmatrix} A \to B \\ B \to A \end{pmatrix},$$

the following equation:

$$\langle\langle B,A\rangle\rangle_\Omega = \frac{1}{\hbar}\int\limits_{-\infty}^{+\infty} J_{A,B}(\nu)\frac{1-e^{-\beta\nu\hbar}}{\Omega-\nu}\,d\nu = \frac{1}{\hbar}\int\limits_{-\infty}^{+\infty} J_{B,A}(-\nu)\frac{1-e^{\beta\nu\hbar}}{\Omega+\nu}\,d\nu$$

$$= \frac{1}{\hbar}\int\limits_{-\infty}^{+\infty} J_{A,B}(\nu)e^{-\beta\nu\hbar}\left(\frac{1-e^{\beta\nu\hbar}}{\Omega+\nu}\right)\,d\nu = -\frac{1}{\hbar}\int\limits_{-\infty}^{+\infty} J_{A,B}(\nu)\frac{1-e^{-\beta\nu\hbar}}{\Omega+\nu}\,d\nu.$$

Thus we have proved the property

$$\langle\langle B,A\rangle\rangle_\Omega = \langle\langle A,B\rangle\rangle_{-\Omega}, \qquad \text{Im}\,\Omega \neq 0. \tag{1.14}$$

1.3. Two-time Correlation Functions and Green Functions for the Linear Polaron Model

Starting from (1.5) and (1.13), we construct a system of equations for Green functions in the case of the linear polaron model[d]:

$$-im\Omega\langle\langle r_\alpha, r_\beta\rangle\rangle_\Omega = \langle\langle p_\alpha, r_\beta\rangle\rangle_\Omega, \tag{1.15}$$

---

[c] Clarification:. Let us write the equality

$$\langle A(t)B(\tau)\rangle_{\text{eq}} = \int\limits_{-\infty}^{+\infty} J_{A,B}(\omega)e^{-i\omega(t-\tau)}\,d\omega.$$

Making the substitutions

$$A \to B, \quad t \to \tau, \quad B \to A, \quad \tau \to t,$$

we arrive at the following result:

$$\langle B(\tau)A(t)\rangle_{\text{eq}} = \int\limits_{-\infty}^{+\infty} J_{B,A}(\omega)e^{-i\omega(\tau-t)}\,d\omega = \int\limits_{-\infty}^{+\infty} J_{B,A}(-\omega)e^{-i\omega(t-\tau)}\,d\omega.$$

[d] Here: $\mathbf{r} = (r_1, r_2, r_3)$, $\mathbf{p} = (p_1, p_2, p_3)$ and $p_\alpha r_\beta - r_\beta p_\alpha = -i\hbar\delta_{\alpha\beta}$.

$$-i\Omega\langle\langle p_\alpha, r_\beta\rangle\rangle_\Omega = -\delta_{\alpha,\beta} - (K_0^2 + \eta^2)\langle\langle r_\alpha, r_\beta\rangle\rangle_\Omega$$

$$- \frac{i}{V^{1/2}}\sum_f S(f)\left(\frac{\hbar}{2\nu(f)}\right)^{1/2} f_\alpha\langle\langle b_f + b_{-f}^\dagger, r_\beta\rangle\rangle_\Omega, \quad (1.16)$$

$$\hbar\Omega\langle\langle b_f, r_\beta\rangle\rangle_\Omega = \hbar\nu(f)\langle\langle b_f, r_\beta\rangle\rangle_\Omega - \frac{i}{V^{1/2}}\left(\frac{\hbar}{2\nu(f)}\right)^{1/2} S(f)\langle\langle \mathbf{f}\cdot\mathbf{r}, r_\beta\rangle\rangle_\Omega,$$

$$\hbar\Omega\langle\langle b_{-f}, r_\beta\rangle\rangle_\Omega = -\hbar\nu(f)\langle\langle b_{-f}^\dagger, r_\beta\rangle\rangle_\Omega \qquad (1.17)$$

$$+ \frac{i}{V^{1/2}}\left(\frac{\hbar}{2\nu(f)}\right)^{1/2} S(f)\langle\langle \mathbf{f}\cdot\mathbf{r}, r_\beta\rangle\rangle_\Omega.$$

From (1.17), we have

$$\langle\langle b_f, r_\beta\rangle\rangle_\Omega = -\frac{i}{V^{1/2}}\frac{1}{\Omega - \nu(f)}\left(\frac{1}{2\hbar\nu(f)}\right)^{1/2} S(f)\langle\langle \mathbf{f}\cdot\mathbf{r}, r_\beta\rangle\rangle_\Omega,$$

$$\langle\langle b_{-f}^\dagger, r_\beta\rangle\rangle_\Omega = \frac{i}{\sqrt{V}}\frac{1}{\Omega + \nu(f)}\left(\frac{1}{2\hbar\nu(f)}\right)^{1/2} S(f)\langle\langle \mathbf{f}\cdot\mathbf{r}, r_\beta\rangle\rangle_\Omega.$$

Thus

$$\langle\langle b_f + b_{-f}^\dagger, r_\beta\rangle\rangle_\Omega$$

$$= \frac{i}{\sqrt{V}}\left(\frac{1}{\Omega + \nu(f)} - \frac{1}{\Omega - \nu(f)}\right)\left(\frac{1}{2\hbar\nu(f)}\right)^{1/2} S(f)\langle\langle \mathbf{f}\cdot\mathbf{r}, r_\beta\rangle\rangle_\Omega.$$

Inserting this formula into (1.15) and (1.16), we find that

$$-m\Omega^2\langle\langle r_\alpha, r_\beta\rangle\rangle_\Omega = -(K_0 + \eta^2)\langle\langle r_\alpha, r_\beta\rangle\rangle_\Omega - \delta_{\alpha,\beta}$$

$$+ \frac{1}{V}\sum_f \frac{S^2(f)}{2\nu(f)} f_\alpha\left(\frac{1}{\Omega + \nu(f)} - \frac{1}{\Omega - \nu(f)}\right)\langle\langle \mathbf{f}\cdot\mathbf{r}, r_\beta\rangle\rangle_\Omega.$$

Since

$$\sum_f F(|f|)f_\alpha f_\beta = \delta_{\alpha\beta}\frac{1}{3}\sum_f F(|f|)f^2$$

and

$$K_0^2 = \frac{1}{3V}\sum_f \frac{S^2(f)f^2}{\nu^2(f)},$$

the following equation results:

$$-m\Omega^2\langle\langle r_\alpha, r_\beta\rangle\rangle_\Omega = -\eta^2\langle\langle r_\alpha, r_\beta\rangle\rangle_\Omega - \frac{1}{3V}\sum_f \frac{S^2(f)f^2}{\nu^2(f)}\langle\langle r_\alpha, r_\beta\rangle\rangle_\Omega$$

$$+ \frac{1}{V} \sum_f \frac{S^2(f)}{6\nu(f)} f^2 \left(\frac{1}{\Omega + \nu(f)} - \frac{1}{\Omega - \nu(f)} \right) \langle\langle r_\alpha, r_\beta \rangle\rangle_\Omega - \delta_{\alpha\beta}. \quad (1.18)$$

However,

$$\frac{1}{\Omega + \nu(f)} - \frac{1}{\nu(f)} + \frac{1}{\nu(f) - \Omega} - \frac{1}{\nu(f)}$$

$$= -\Omega \left(\frac{1}{\nu(f)\{\Omega + \nu(f)\}} + \frac{1}{\{\Omega - \nu(f)\}\nu(f)} \right).$$

Taking account of this transformation, (1.18) can be represented as

$$\delta_{\alpha,\beta} = \langle\langle r_\alpha, r_\beta \rangle\rangle_\Omega$$

$$\times \left\{ m\Omega^2 - \eta^2 - \frac{\Omega}{V} \sum_{(f)} \frac{S^2(f)}{6\nu^2(f)} f^2 \left(\frac{1}{\Omega + \nu(f)} + \frac{1}{\Omega - \nu(f)} \right) \right\}. \quad (1.19)$$

Let us define

$$\triangle(\Omega) = -\frac{1}{V} \sum_{(f)} \frac{S^2(f)}{6\nu^2(f)} f^2 \left(\frac{1}{\Omega + \nu(f)} + \frac{1}{\Omega - \nu(f)} \right) \quad (1.20)$$

and note that

$$\triangle(-\Omega) = -\triangle(\Omega). \quad (1.21)$$

Then (1.19) can be rewritten in the form

$$\langle\langle r_\alpha, r_\beta \rangle\rangle_\Omega = \frac{\delta_{\alpha,\beta}}{m\Omega^2 - \eta^2 + \Omega\triangle(\Omega)}. \quad (1.22)$$

Taking (1.15) into account, we have also

$$\langle\langle p_\alpha, r_\beta \rangle\rangle_\Omega = \frac{-im\Omega\delta_{\alpha,\beta}}{m\Omega^2 - \eta^2 + \Omega\triangle(\Omega)}. \quad (1.23)$$

Recalling (1.14), we get

$$\langle\langle r_\beta, p_\alpha \rangle\rangle_\Omega = \frac{im\Omega\delta_{\alpha,\beta}}{m\Omega^2 - \eta^2 + \Omega\triangle(\Omega)}. \quad (1.24)$$

Using (1.5) and (1.13), we have further

$$-i\Omega m\langle\langle r_\beta, p_\alpha\rangle\rangle_\Omega = \frac{m}{i\hbar}\langle r_\beta p_\alpha - p_\alpha r_\beta\rangle + \langle\langle p_\beta, p_\alpha\rangle\rangle_\Omega,$$

so that

$$\langle\langle p_\beta, p_\alpha\rangle\rangle_\Omega = -m\delta_{\alpha,\beta} + \frac{(m\Omega)^2\delta_{\alpha,\beta}}{m\Omega^2 - \eta^2 + \Omega\triangle(\Omega)} = -\delta_{\alpha,\beta}\frac{m(\Omega\triangle(\Omega) - \eta^2)}{m\Omega^2 - \eta^2 + \Omega\triangle(\Omega)}.$$

$$(1.25)$$

For $|\text{Im}\,\Omega| > 0$ we can take the limit $\eta \to 0$, in (1.22) to (1.25). Then

$$\langle\langle r_\alpha, r_\beta\rangle\rangle_\Omega = \frac{\delta_{\alpha,\beta}}{m\Omega^2 + \Omega\triangle(\Omega)},$$

$$\langle\langle p_\alpha, r_\beta\rangle\rangle_\Omega = -\langle\langle r_\beta, p_\alpha\rangle\rangle_\Omega = -\frac{im\delta_{\alpha,\beta}}{m\Omega + \triangle(\Omega)}, \qquad (1.26)$$

$$\langle\langle p_\alpha, p_\beta\rangle\rangle_\Omega = -\delta_{\alpha,\beta}\frac{m\triangle(\Omega)}{m\Omega + \triangle(\Omega)}.$$

It should be stressed that when calculating a spectral intensity $J_{A,B}(\omega)$, for example $J_{p_\alpha,p_\beta}(\omega)$, we have to use (1.25), which contains $\eta > 0$, and only after this can we take the limit $\eta \to 0$.

By means of (1.11), we arrive at the following spectral density:

$$J_{p_\alpha,p_\beta}(\omega) = \delta_{\alpha,\beta}\frac{i\hbar}{2\pi}\left(1 - e^{-\beta\hbar\omega}\right)^{-1}\frac{(m\Omega)^2}{m\Omega^2 - \eta^2 + \Omega\triangle(\Omega)}\Bigg|_{\omega-i0}^{\omega+i0}$$

$$= -m\delta_{\alpha,\beta}\frac{i\hbar}{2\pi(1 - e^{-\beta\hbar\omega})}\left(\frac{\Omega\triangle(\Omega) - \eta^2}{m\Omega^2 - \eta^2 + \Omega\triangle(\Omega)}\right)\Bigg|_{\omega-i0}^{\omega+i0}. \quad (1.27)$$

Here we have introduced the notation

$$F(\Omega)\Big|_a^b = F(b) - F(a).$$

It must be kept in mind that division by $(1 - e^{-\beta\hbar\omega})^{-1}$ in (1.27) may lead to a delta function $K\delta(\omega)$ with some unknown coefficient K. However, when $\eta^2 > 0$, the expression

$$\cdots\Big|_{\omega-i0}^{\omega+i0}$$

in (1.27) is equal to zero in the vicinity of the point $\omega = 0$. On the other hand, we know that in this case (namely for $\eta^2 > 0$), function $p_\alpha(t)$ can be represented as a sum of harmonic oscillations with nonzero frequencies. Therefore the corresponding spectral intensity

$$J_{p_\alpha p_\beta}(\omega) = 0$$

in a neighborhood of the point $\omega = 0$ that does not contain the harmonic-oscillator frequencies. Therefore we have to calculate (1.27) first of all under the condition $\eta^2 > 0$, and only after this can we take the limit $\eta \to 0$.

Let us consider in detail the simplest example when

$$\nu(f) = \nu = \text{const} > 0. \tag{1.28}$$

In this case (1.20) if one takes into account that

$$K_0^2 = \frac{1}{3V} \sum_{(f)} \frac{S^2(f)}{\nu^2(f)} f^2$$

can be rewritten as

$$\triangle(\Omega) = -\frac{K_0^2}{2} \left(\frac{1}{\Omega + \nu} + \frac{1}{\Omega - \nu} \right) = -\frac{K_0^2 \Omega}{\Omega^2 - \nu^2}. \tag{1.29}$$

From (1.27), we derive

$$J_{p_\alpha p_\beta}(\omega) = \delta_{\alpha,\beta} \frac{i\hbar}{2\pi} \left(1 - e^{-\beta\hbar\omega} \right)^{-1} \frac{(m\Omega)^2(\Omega^2 - \nu^2)}{m\Omega^4 - \Omega^2(K_0^2 + \eta^2 + m\nu^2) + \nu^2\eta^2} \Big|_{\omega-i0}^{\omega+i0}. \tag{1.30}$$

Here the denominator has two roots with respect to Ω^2:

$$\omega_1^2 = \frac{\nu^2\eta^2}{K_0^2 + \nu^2 m} + O(\eta^4), \qquad \omega_2^2 = \frac{K_0^2 + \nu^2 m}{m} + O(\eta^2). \tag{1.31}$$

Hence

$$m\Omega^4 - \Omega^2(K_0^2 + \eta^2 + \nu^2 m) + \nu^2\eta^2 = m(\Omega^2 - \omega_1^2)(\Omega^2 - \omega_2^2).$$

Therefore

$$\frac{m^2\Omega^2(\Omega^2 - \nu^2)}{m\Omega^4 - \Omega^2(K_0^2 + \eta^2 + \nu^2 m) + \nu^2\eta^2}$$

$$= \frac{m\omega_1^2(\omega_1^2 - \nu^2)}{\omega_1^2 - \omega_2^2} \frac{1}{\Omega^2 - \omega_1^2} + \frac{m\omega_2^2(\omega_2^2 - \nu^2)}{\omega_2^2 - \omega_1^2} \frac{1}{\Omega^2 - \omega_2^2} + \mathcal{E},$$

where the expression $\mathcal{E}$ is regular and does not have singularities on the real axis. On the other hand,

$$\frac{1}{\Omega^2 - \omega_j^2} = \frac{1}{2\omega_j} \left(\frac{1}{\Omega - \omega_j} - \frac{1}{\Omega + \omega_j} \right), \quad j = 1,2,$$

and

$$\frac{1}{\Omega^2 - \omega_j^2} \Big|_{\omega-i0}^{\omega+i0} = \frac{\pi i}{\omega_j} \left\{ \delta(\omega - \omega_j) - \delta(\omega + \omega_j) \right\}.$$

Hence it follows from (1.30) that

$$
\begin{aligned}
J_{p_\alpha p_\beta}(\omega) \\
= \frac{1}{2} \delta_{\alpha,\beta} \frac{m(\nu^2 - \omega_1^2)}{\omega_2^2 - \omega_1^2} \left(\frac{\hbar\omega_1}{1 - e^{-\beta\hbar\omega_1}} \delta(\omega - \omega_1) + \frac{\hbar\omega_1}{e^{\beta\hbar\omega_1} - 1} \delta(\omega + \omega_1) \right) \\
+ \frac{1}{2} \delta_{\alpha,\beta} \frac{m(\omega_2^2 - \nu^2)}{\omega_2^2 - \omega_1^2} \left(\frac{\hbar\omega_2}{1 - e^{-\beta\hbar\omega_2}} \delta(\omega - \omega_2) + \frac{\hbar\omega_2}{e^{\beta\hbar\omega_2} - 1} \delta(\omega + \omega_2) \right).
\end{aligned}
$$

$$(1.32)$$

Keeping in mind (1.31), one has $\omega_1 \to 0$, when $\eta \to 0$, and at the same time

$$
\frac{\hbar\omega_1}{1 - e^{-\beta\hbar\omega_1}} \to \frac{1}{\beta} = \vartheta, \qquad \frac{\hbar\omega_1}{e^{\beta\hbar\omega_1} - 1} \to \frac{1}{\beta} = \vartheta,
$$

$$
\omega_2 \to \mu = \left(\frac{K_0^2}{m} + \nu^2 \right)^{1/2}.
$$

Taking the limit $\eta \to 0$ in (1.32), we find that

$$
J_{p_\alpha p_\beta}(\omega) = \delta_{\alpha,\beta} \frac{m^2 \nu^2 \vartheta}{K_0^2 + m\nu^2} \delta(\omega)
$$

$$
+ \frac{K_0^2 \delta_{\alpha,\beta}}{2\mu} \left(\frac{\hbar}{1 - e^{-\beta\hbar\mu}} \delta(\omega - \mu) + \frac{\hbar}{e^{\beta\hbar\mu} - 1} \delta(\omega + \mu) \right). \quad (1.33)
$$

If one knows the spectral intensity, one can easily calculate two-time correlation functions:

$$
\langle p_\alpha(t) p_\beta(\tau) \rangle_{\text{eq}} = 0, \quad \alpha \neq \beta,
$$

$$
\langle p_\alpha(t) p_\alpha(\tau) \rangle_{\text{eq}} \tag{1.34}
$$

$$
= \frac{m^2 \nu^2 \vartheta}{K_0^2 + m\nu^2} + \frac{K_0^2}{2\mu} \left(\frac{\hbar}{1 - e^{-\beta\hbar\mu}} e^{-i\mu(t-\tau)} + \frac{\hbar}{e^{\beta\hbar\mu} - 1} e^{i\mu(t-\tau)} \right).
$$

Now we consider the more general case when $\nu(f)$ possesses a continuous spectrum in the limit $V \to \infty$. Let us return to (1.27) and transform this formula into a new one:

$$
J_{p_\alpha p_\beta}(\omega) = m\delta_{\alpha,\beta} \frac{i\hbar\omega}{2\pi(1 - e^{-\beta\hbar\omega})} \frac{1}{\omega} f_\eta(\Omega) \Big|_{\omega - i0}^{\omega + i0}, \tag{1.35}
$$

where

$$
-f_\eta(\Omega) = \frac{\Omega\triangle(\Omega) - \eta^2}{m\Omega^2 - \eta^2 + \Omega\triangle(\Omega)}.
$$

We can see that

$$
f_\eta(0) = -1 \quad \text{and} \quad f_\eta(\Omega) \Big|_{\omega - i0}^{\omega + i0} = 0
$$

for small enough ω. Hence, for small enough ω

$$\frac{1}{\Omega} f_\eta(\Omega) \Big|_{\omega-i0}^{\omega+i0} = f_\eta(\omega) \left(\frac{1}{\omega+i\varepsilon} - \frac{1}{\omega-i\varepsilon} \right) = -2\pi i f_\eta(\omega)\delta(\omega) = 2\pi i\delta(\omega).$$
(1.36)

It should be stressed that the function $1/\Omega$ has only one singular point $\Omega = 0$. So, for arbitrary real ω

$$\frac{1}{\Omega} f_\eta(\Omega) \Big|_{\omega-i0}^{\omega+i0} = 2\pi i\delta(\omega) + \frac{1}{\omega} f_\eta(\Omega) \Big|_{\omega-i0}^{\omega+i0}$$

or

$$\frac{1}{\omega} f_\eta(\Omega) \Big|_{\omega-i0}^{\omega+i0} = -2\pi i\delta(\omega) + \frac{1}{\Omega} f_\eta(\Omega) \Big|_{\omega-i0}^{\omega+i0}.$$

Therefore it follows from (1.35) that

$$J_{p_\alpha p_\beta}(\omega) = \delta_{\alpha,\beta} \frac{i\hbar\omega m}{2\pi(1-e^{-\beta\hbar\omega})} \left(-2\pi i\delta(\omega) + \frac{1}{\Omega} f_\eta(\Omega) \Big|_{\omega-i0}^{\omega+i0} \right).$$

But the function

$$\frac{\Omega}{1-e^{-\beta\hbar\Omega}}$$

is regular in the vicinity of the real axis, and so

$$\frac{\hbar\omega}{1-e^{-\beta\hbar\omega}} \frac{1}{\Omega} f_\eta(\Omega) \Big|_{\omega-i0}^{\omega+i0} = \frac{\hbar}{1-e^{-\beta\hbar\Omega}} f_\eta(\Omega) \Big|_{\omega-i0}^{\omega+i0}.$$

Thus

$$J_{p_\alpha p_\beta}(\omega) = 0 \quad \text{if} \quad \alpha \neq \beta,$$

and

$$J_{p_\alpha p_\alpha}(\omega) = m\vartheta\delta(\omega) - \frac{i\hbar m}{2\pi(1-e^{-\beta\hbar\Omega})} \frac{\Omega\triangle(\Omega)-\eta^2}{m\Omega^2-\eta^2+\Omega\triangle(\Omega)} \Big|_{\omega-i0}^{\omega+i0}.$$

As a result, we have

$$\langle p_\alpha(t)p_\beta(\tau)\rangle_{\text{eq}} = 0 \quad \text{for} \quad \alpha \neq \beta,$$

$$\langle p_\alpha(t)p_\alpha(\tau)\rangle_{\text{eq}} = m\vartheta - \int_{-\infty}^{+\infty} \frac{i\hbar m e^{-i\Omega(t-\tau)}}{2\pi(1-e^{-\beta\hbar\Omega})} \frac{\Omega\triangle(\Omega)-\eta^2}{m\Omega^2-\eta^2+\Omega\triangle(\Omega)} \Big|_{\omega-i0}^{\omega+i0} d\omega.$$

Consider the function

$$f_\eta(\Omega) = -\frac{\Omega\triangle(\Omega)-\eta^2}{m\Omega^2-\eta^2+\Omega\triangle(\Omega)} = -1 + \frac{m\Omega^2}{m\Omega^2-\eta^2+\Omega\triangle(\Omega)}$$

$$= -1 + \frac{m\Omega}{m\Omega+\triangle(\Omega)-\eta^2/\Omega}$$

for $\Omega = i\varepsilon + \omega$ and $\Omega = -i\varepsilon + \omega$; $\varepsilon > 0$. Taking account of (1.20), we have

$$-\mathrm{Im}\,\triangle(\omega + \varepsilon)$$

$$= \frac{\varepsilon}{V} \sum_{(f)} \frac{S^2(f)}{6\nu^2(f)} f^2 \left(\frac{1}{\{\nu(f) + \omega\}^2 + \varepsilon^2} + \frac{1}{\{\nu(f) - \omega\}^2 + \varepsilon^2} \right) > 0.$$

Then it is obvious that

$$\mathrm{Im}\,\frac{-\eta^2}{\Omega} = +\frac{\eta^2 \varepsilon}{\omega^2 + \varepsilon^2} > 0, \quad \Omega = \omega + i\varepsilon.$$

Hence

$$\mathrm{Im}\left(m\Omega + \triangle(\Omega) - \frac{\eta^2}{\Omega} \right) > \varepsilon m, \quad \Omega = \omega + i\varepsilon$$

and

$$\left| m\Omega + \triangle(\Omega) - \frac{\eta^2}{\Omega} \right| > \varepsilon m. \tag{1.37}$$

In the same way, it can be proved that, for $\varepsilon = -\mathrm{Im}\,\Omega > 0$

$$\left| m\Omega + \triangle(\Omega) - \frac{\eta^2}{\Omega} \right| > \varepsilon.$$

Therefore the function $f_\eta(\Omega)$ is a regular function of the complex variable Ω on the two half-planes

$$\mathrm{Im}\,\Omega > 0 \quad \text{and} \quad \mathrm{Im}\,\Omega < 0. \tag{1.38}$$

Then we note that the poles of the function

$$\frac{1}{1 - e^{-\beta\hbar\Omega}}$$

in the domain (1.38) in the vicinity of the real axis are

$$\Omega = \frac{2\pi i}{\hbar\beta} \quad \text{and} \quad \Omega = -\frac{2\pi i}{\hbar\beta}.$$

in the domain (1.38) in the vicinity of the real axis are

$$0 < \mathrm{Im}\,\Omega < \frac{2\pi}{\hbar\beta}$$

or in the region

$$0 > \mathrm{Im}\,\Omega > -\frac{2\pi}{\hbar\beta}.$$

Then

$$\int_{\mathcal{L}} \frac{e^{-i\Omega(t-\tau)}}{1 - e^{-\beta\hbar\Omega}} f_\eta(\Omega)\,d\Omega = 0.$$

We recall (1.20),

$$\triangle(\Omega) = -\frac{1}{V} \sum_{(f)} \frac{S^2(f)}{6\nu^2(f)} f^2 \left(\frac{1}{\nu(f) + \Omega} + \frac{1}{\Omega - \nu(f)} \right), \qquad (1.39)$$

which contains only a finite number (f), of terms if the volume V is fixed. Therefore, for large enough $|\Omega|$,

$$|\Omega \triangle(\Omega)| = \text{const}, \qquad |\text{Im}\, \Omega| \geqslant \varepsilon > 0 \qquad (1.40)$$

$$|\triangle(\Omega)| \leqslant \frac{\text{const}}{|\Omega|}.$$

As a result, one can choose for the contour $\mathcal{L}$ an infinite contour (see Fig. 1):

$$i\varepsilon_1 - \infty < \overrightarrow{\omega} < i\varepsilon_1 + \infty, \qquad i\varepsilon - \infty < \overleftarrow{\omega} < i\varepsilon + \infty,$$

$$0 < \varepsilon < \varepsilon_1 < \frac{2\pi}{\hbar\beta}$$

Fig. 1.1.

and obtain the expression

$$-\frac{im\hbar}{2\pi} \int\limits_{i\varepsilon_1 - \infty}^{i\varepsilon_1 + \infty} \frac{e^{-i\Omega(t-\tau)}}{1 - e^{-\beta\hbar\Omega}} \frac{\Omega\triangle(\Omega) - \eta^2}{m\Omega^2 - \eta^2 + \Omega\triangle(\Omega)} d\Omega$$

$$= +\frac{im\hbar}{2\pi} \int\limits_{i\varepsilon - \infty}^{i\varepsilon + \infty} \frac{e^{-i\Omega(t-\tau)}}{1 - e^{-\beta\hbar\Omega}} \frac{\Omega\triangle(\Omega) - \eta^2}{m\Omega^2 - \eta^2 + \Omega\triangle(\Omega)} d\Omega = 0,$$

from which it follows that the integral

$$-\frac{im\hbar}{2\pi} \int\limits_{i\varepsilon - \infty}^{i\varepsilon + \infty} \frac{e^{-i\Omega(t-\tau)}}{1 - e^{-\beta\hbar\Omega}} \frac{\Omega\triangle(\Omega) - \eta^2}{m\Omega^2 - \eta^2 + \Omega\triangle(\Omega)} d\Omega$$

does not depend on the magnitude of ε, when ε belongs to the domain

$$0 < \varepsilon < \frac{2\pi}{\hbar\beta}. \qquad (1.41)$$

In full analogy with the previous consideration, one can prove that the same integral, calculated along the closed contour $(-i\varepsilon - \infty, -i\varepsilon + \infty)$, situated on the lower half-plane, does not depend on the precise value of ε, if ε belongs to the region (1.41). We can write (1.36) in the form

$$\langle p_\alpha(t)p_\alpha(\tau)\rangle_{eq} = m\vartheta - \frac{im\hbar}{2\pi} \int\limits_{i\varepsilon-\infty}^{i\varepsilon+\infty} \frac{e^{-i\Omega(t-\tau)}}{1 - e^{-\beta\hbar\Omega}} \frac{\Omega\triangle(\Omega) - \eta^2}{m\Omega^2 - \eta^2 + \Omega\triangle(\Omega)} d\Omega$$

$$+ \frac{im\hbar}{2\pi} \int\limits_{-i\varepsilon-\infty}^{-i\varepsilon+\infty} \frac{e^{-i\Omega(t-\tau)}}{1 - e^{-\beta\hbar\Omega}} \frac{\Omega\triangle(\Omega) - \eta^2}{m\Omega^2 - \eta^2 + \Omega\triangle(\Omega)} d\Omega.$$

Here, owing to (1.37) and (1.40), we may pass to the limit $\eta \to 0$ and write

$$\langle p_\alpha(t)p_\alpha(\tau)\rangle_{eq} = m\vartheta - \frac{im\hbar}{2\pi} \int\limits_{i\varepsilon-\infty}^{i\varepsilon+\infty} \frac{e^{-\Omega(t-\tau)}}{1 - e^{-\beta\hbar\Omega}} \frac{\triangle(\Omega)}{m\Omega + \triangle(\Omega)} d\Omega$$

$$+ \frac{im\hbar}{2\pi} \int\limits_{-i\varepsilon-\infty}^{-i\varepsilon+\infty} \frac{e^{-i\Omega(t-\tau)}}{1 - e^{-\beta\hbar\Omega}} \frac{\triangle(\Omega)}{m\Omega + \triangle(\Omega)} d\Omega. \quad (1.42)$$

Consider now the standard limit $V \to \infty$. It follows from (1.20) that

$$\triangle(\Omega) = - \int\limits_{-\infty}^{+\infty} E_V(\nu)\frac{d\nu}{\Omega - \nu},$$

where

$$E_V(\omega) = \frac{1}{V} \sum_{(f)} \frac{S^2(f)}{6\nu^2(f)} f^2 \left\{ \delta(\nu(f) + \omega) + \delta(\nu(f) - \omega) \right\},$$

$$E_V \geqslant 0, \quad E_V(-\omega) = E_V(\omega).$$

Let us suppose that this generalized function $E_V(\omega)$ has the behavior

$$E_V(\omega) \to E(\omega) = \frac{1}{(2\pi)^3} \int \frac{S^2(f)}{6\nu^2(f)} f^2 \left\{ \delta(\nu(f) + \omega) + \delta(\nu(f) - \omega) \right\} d\mathbf{f}$$

if $V \to \infty$ in such a way that the convergence of the function:
(1°) is uniform on any finite interval

$$+i\varepsilon - \infty < \omega < i\varepsilon + \infty, \quad \varepsilon > 0 \qquad (1.43)$$

(on the upper half-plane) and

$$-i\varepsilon - \infty < \omega < -i\varepsilon + \infty, \quad \varepsilon > 0 \qquad (1.44)$$

(on the lower half-plane):

$$\int\limits_{-\infty}^{+\infty} E_V(\nu)\frac{d\nu}{\Omega - \nu} \to -\triangle_\infty(\Omega) = \int\limits_{-\infty}^{+\infty} \frac{E(\nu)\,d\nu}{\Omega - \nu};$$

($2°$) for $|\operatorname{Im}\Omega| \geqslant \varepsilon$

$$|\Omega\triangle(\Omega)| \leqslant K_\varepsilon,$$

where K_ε — is some constant independent of the volume V.

Under these conditions, we can go to the limit $V \to \infty$ in (1.42) and write

$$\langle p_\alpha(t)p_\alpha(\tau)\rangle_{\mathrm{eq}} = m\vartheta - \frac{im\hbar}{2\pi} \int_{i\varepsilon-\infty}^{i\varepsilon+\infty} \frac{e^{-i\Omega(t-\tau)}}{1-e^{-\beta\hbar\Omega}} \frac{\triangle_\infty(\Omega)}{m\Omega + \triangle_\infty(\Omega)}\, d\Omega +$$

$$+ \frac{im\hbar}{2\pi} \int_{-i\varepsilon-\infty}^{-i\varepsilon+\infty} \frac{e^{-i\Omega(t-\tau)}}{1-e^{-\beta\hbar\Omega}} \frac{\triangle_\infty(\Omega)}{m\Omega + \triangle_\infty(\Omega)}\, d\Omega =$$

$$= m\vartheta - \frac{im\hbar}{2\pi} \int_{i\varepsilon-\infty}^{i\varepsilon+\infty} \frac{e^{-i\Omega(t-\tau)}}{1-e^{-\beta\hbar\Omega}} \frac{\triangle_\infty(\Omega)}{m\Omega + \triangle_\infty(\Omega)}\, d\omega\bigg|_{\omega-i\varepsilon}^{\omega+i\varepsilon}, \quad (1.45)$$

where

$$\triangle(\Omega) = - \int_{-\infty}^{+\infty} \frac{E(\nu)}{\Omega-\nu}\, d\nu, \qquad E(\nu) = E(-\nu) \geqslant 0, \quad (1.46)$$

and, from ($2°$)

$$|\Omega\triangle_\infty(\Omega)| \leqslant K_\varepsilon \quad \text{for} \quad |\operatorname{Im}\Omega| \geqslant \varepsilon. \quad (1.47)$$

The right-hand side of (1.45) is independent of the value of ε, if ε is positive and small enough. Owing to this fact, we can take the passage to the limit $\varepsilon \to 0$, $\varepsilon > 0$ in (1.45). The result is [e]

$$\langle p_\alpha(t)p_\alpha(\tau)\rangle_{\mathrm{eq}} = m\vartheta - \frac{im\hbar}{2\pi} \int_{-\infty}^{+\infty} \frac{\omega e^{-i\omega(t-\tau)}}{1-e^{-\beta\hbar\omega}} \frac{1}{\Omega}\frac{\triangle_\infty(\Omega)}{m\Omega + \triangle_\infty(\Omega)}\, d\omega\bigg|_{\omega-i0}^{\omega+i0}.$$

Since

$$-\frac{\triangle_\infty(\Omega)}{m\Omega + \triangle_\infty(\Omega)} = -1 + \frac{m\Omega}{m\Omega + \triangle_\infty(\Omega)}$$

and

$$-\frac{i\hbar m}{2\pi}\frac{\omega}{1-e^{-\beta\hbar\omega}}\frac{1}{\Omega}\bigg|_{\omega-i0}^{\omega+i0} = -\frac{m}{\beta}\,\delta(\omega) = -m\vartheta\delta(\omega),$$

we come at last to the expression

$$\langle p_\alpha(t)p_\alpha(\tau)\rangle_{\mathrm{eq}} = \int_{-\infty}^{+\infty} J(\omega)e^{-i\omega(t-\tau)}\, d\omega, \quad (1.48)$$

---

[e] Here we assume that discontinuities of the expression

$$\frac{1}{\Omega}\frac{\triangle_\infty(\Omega)}{m\Omega + \triangle_\infty(\Omega)}\bigg|_{\omega-i0}^{\omega+i0}$$

are of the first order, so that if $F(\Omega)$ — is an analytic function in the vicinity of the real axis, such as the function

$$F(\Omega) = \frac{\Omega}{1-e^{-\beta\hbar\omega}}, \quad \text{then} \quad \{F(\Omega) - F(\omega)\}\frac{1}{\Omega}\frac{\triangle_\infty(\Omega)}{m\Omega + \triangle_\infty(\Omega)}\bigg|_{\omega-i0}^{\omega+i0} = 0.$$

where
$$J(\omega) = \frac{i\hbar m^2}{2\pi} \frac{\omega}{1 - e^{-\beta\hbar\omega}} \frac{1}{m\Omega + \triangle_\infty(\Omega)}\bigg|_{\omega-i0}^{\omega+i0}.$$

From (1.46), one can derive
$$\triangle_\infty(\omega \pm i0) = -\int E(\nu)\mathcal{P}\left(\frac{1}{\omega - \nu}\right) d\nu \pm i\pi E(\omega).$$

It follows from this equation that
$$i\frac{1}{m\Omega + \triangle_\infty(\Omega)}\bigg|_{\omega-i0}^{\omega+i0} \geqslant 0.$$

Because, in addition,
$$\frac{\omega}{1 - e^{-\beta\hbar\omega}} \geqslant 0,$$

it is easy to find that
$$J(\omega) \geqslant 0. \tag{1.49}$$

It should be noted that if we put $E(\omega)$ equal to
$$E(\omega) = \frac{K_0^2}{2}\left\{\delta(\omega - \nu_0) + \delta(\omega + \nu_0)\right\},$$

we get from (1.48) our previous (1.34) that we derived earlier.

1.4. Free Energy Calculation for the Linear Polaron Model

We now proceed with the calculation of the free energy for the dynamical system under consideration. The free energy is defined as
$$F = -\vartheta \ln \operatorname{Tr} e^{\beta H}, \qquad (H = H_{\text{linear model}}).$$

The free energy for a single free particle of mass m that does not interact with the phonon bath is
$$F_s = -\vartheta \ln \operatorname{Tr} \exp\left(-\frac{\mathbf{p}^2\beta}{2m}\right) = -\vartheta \lim_{\eta\to 0} \ln \operatorname{Tr} \exp\left\{-\left(\frac{\mathbf{p}^2}{2m} + \frac{\eta^2}{2}\mathbf{r}^2\right)\beta\right\}.$$

The free energy of the free-phonon field Σ is
$$F_\Sigma = -\vartheta \ln \operatorname{Tr} e^{-H_\Sigma\beta}, \qquad H_\Sigma = \frac{1}{2}\sum_{(f)}\{p_f p_f^* + \nu(f)^2 q_f q_f^*\}.$$

Because explicit expressions for these energies are well known, we need calculate only that part of the total free energy that is due to the interaction between the phonon field and the particle, i.e.
$$F_{\text{int}} = F - F_S - F_\Sigma.$$

For technical reasons, we introduce an auxiliary parameter λ $(0 \leqslant \lambda \leqslant 1)$ into the total Hamiltonian $H = H_{\text{linear model}}$:

$$H(\lambda) = \frac{\mathbf{p}^2}{2m} + \frac{\eta^2 \mathbf{r}^2}{2} + \frac{\lambda^2}{V} \sum_{(f)} \frac{S^2(f)}{6\nu^2(f)} f^2 \mathbf{r}^2 + \frac{i}{V^{1/2}} \sum_{(f)} \lambda S_f q_f \, \mathbf{f} \cdot \mathbf{r} + H_\Sigma.$$

We see that

$$H(0) = H_S + H_\Sigma, \qquad H(1) = H.$$

Thus

$$F_{\text{int}} = \int\limits_0^1 d\lambda \, \frac{\partial F(\lambda)}{\partial \lambda} = -\vartheta \int\limits_0^1 d\lambda \, \frac{\partial}{\partial \lambda} \ln \text{Tr} \, e^{-\beta H(\lambda)} =$$

$$= \int\limits_0^1 d\lambda \, \frac{\text{Tr}[\{\partial H(\lambda)/\partial \lambda\} e^{-\beta H(\lambda)}]}{\text{Tr} \, e^{-\beta H(\lambda)}} = \int\limits_0^1 d\lambda \left\langle \frac{\partial H}{\partial \lambda} \right\rangle_{\lambda,\text{eq}}, \quad (1.50)$$

i.e.

$$F_{\text{int}} = \int\limits_0^1 d\lambda \left\langle \frac{\partial H}{\partial \lambda} \right\rangle_{\lambda,\text{eq}}. \quad (1.51)$$

In this formula the subscript λ indicates that the averaging is with respect to the Hibbs equilibrium statistical operator corresponding to the Hamiltonian $H(\lambda)$:

$$\rho(H(\lambda)) = e^{-\beta H(\lambda)} / \text{Tr}(e^{-\beta H(\lambda)}).$$

But

$$\frac{\partial H(\lambda)}{\partial \lambda} = \frac{\lambda}{3V} \sum_{(f)} \frac{S^2(f)}{\nu^2(f)} f^2 \mathbf{r}^2 + \frac{i}{V^{1/2}} \sum_{(f)} S_f q_f \, \mathbf{f} \cdot \mathbf{r}.$$

From another point of view, we can write the equations of motion (1.5) for the Hamiltonian $H(\lambda)$:

$$-m \frac{d^2 \mathbf{r}}{dt^2} - \eta^2 \mathbf{r} = \frac{\lambda^2}{3} \sum_{(f)} \frac{S^2(f)}{\nu^2(f)} f^2 \mathbf{r} + \frac{i}{\sqrt{V}} \sum_{(f)} \lambda S_f q_f \mathbf{f}.$$

Therefore

$$\lambda \left\langle \frac{\partial H(\lambda)}{\partial \lambda} \right\rangle_{\lambda,\text{eq}} = -\left\langle \left(m \frac{d^2 \mathbf{r}}{dt^2} + \eta^2 \mathbf{r} \right) \cdot \mathbf{r} \right\rangle_{\lambda,\text{eq}}. \quad (1.52)$$

Taking (1.6), (1.11) and (1.22) into account, we have

$$\langle r_\alpha(t) r_\beta(\tau) \rangle_{\lambda,\text{eq}} = 0 \quad \text{for} \quad \alpha \neq \beta,$$

$$\langle r_\alpha(t) r_\alpha(\tau) \rangle_{\lambda,\text{eq}} = \frac{i\hbar}{2\pi} \int\limits_{-\infty}^{+\infty} \frac{e^{-i\omega(t-\tau)}}{1 - e^{-\beta\hbar\omega}} \frac{1}{m\Omega^2 - \eta^2 + \lambda^2 \Omega \triangle(\Omega)} \bigg|_{\omega-i0}^{\omega+i0} d\omega.$$

$$(1.53)$$

Here we have not forgotten that on changing from the Hamiltonian H to the Hamiltonian $H(\lambda)$, we have introduced the parameter λ into S_f, which results in the necessity to change $\triangle(\Omega)$ into $\lambda^2\triangle(\Omega)$. So

$$\left\langle \left(-m\frac{d^2 r_\alpha(t)}{dt^2} - \eta^2 r_\alpha(t)\right) r_\alpha(\tau) \right\rangle_{\lambda,\text{eq}}$$

$$= \frac{i\hbar}{2\pi} \int_{-\infty}^{+\infty} \frac{m\omega^2 - \eta^2}{1 - e^{-\beta\hbar\omega}} e^{-i\omega(t-\tau)} \frac{1}{m\Omega^2 - \eta^2 + \lambda^2\Omega\triangle(\Omega)} \Big|_{\omega-i0}^{\omega+i0} d\omega$$

$$= \frac{i\hbar}{2\pi} \int_{-\infty}^{+\infty} \frac{e^{-i\omega(t-\tau)}}{1 - e^{-\beta\hbar\omega}} \frac{m\Omega^2 - \eta^2}{m\Omega^2 - \eta^2 + \lambda^2\Omega\triangle(\Omega)} \Big|_{\omega-i0}^{\omega+i0} d\omega,$$

if $\Omega \longrightarrow \omega$, and, inserting this result into (1.51) we see that

$$\lambda\left\langle \frac{\partial H(\lambda)}{\partial \lambda} \right\rangle_{\lambda,\text{eq}} = 3\frac{i\hbar}{2\pi} \int_{-\infty}^{+\infty} \frac{1}{1 - e^{-\beta\hbar\omega}} \frac{m\Omega^2 - \eta^2}{m\Omega^2 - \eta^2 + \lambda^2\Omega\triangle(\Omega)} \Big|_{\omega-i0}^{\omega+i0} d\omega.$$

Let us note that

$$\frac{m\Omega^2 - \eta^2}{m\Omega^2 - \eta^2 + \lambda^2\Omega\triangle(\Omega)} = 1 - \frac{\Omega\lambda^2\triangle(\Omega)}{m\Omega^2 - \eta^2 + \lambda^2\Omega\triangle(\Omega)},$$

$$\frac{m\Omega^2 - \eta^2}{m\Omega^2 - \eta^2 + \lambda^2\Omega\triangle(\Omega)} \Big|_{\omega-i0}^{\omega+i0} = -\frac{\Omega\lambda^2\triangle(\Omega)}{m\Omega^2 - \eta^2 + \lambda^2\Omega\triangle(\Omega)} \Big|_{\omega-i0}^{\omega+i0}.$$

Hence (1.51) gives us

$$\left\langle \frac{\partial H(\lambda)}{\partial \lambda} \right\rangle_{\lambda,\text{eq}} = -3\frac{i\hbar}{2\pi} \int_{-\infty}^{+\infty} \frac{1}{1 - e^{-\beta\hbar\omega}} \frac{\Omega\lambda\triangle(\Omega)}{m\Omega^2 - \eta^2 + \lambda^2\Omega\triangle(\Omega)} \Big|_{\omega-i0}^{\omega+i0} d\omega$$

$$= -3\frac{i\hbar}{2\pi} \int_{-\infty}^{+\infty} \frac{\Omega}{1 - e^{-\beta\hbar\Omega}} \frac{\lambda\triangle(\Omega)}{m\Omega^2 - \eta^2 + \lambda^2\Omega\triangle(\Omega)} \Big|_{\omega-i0}^{\omega+i0} d\omega. \quad (1.54)$$

Using a similar approach for the calculation of the correlation function

$$\langle p_\alpha(t) p_\alpha(\tau) \rangle_{\text{eq}},$$

we find in the limit $\eta \to 0$, $V \to \infty$ that

$$\left\langle \frac{\partial H(\lambda)}{\partial \lambda} \right\rangle_{\lambda,\text{eq}} = -\frac{3i\hbar}{2\pi} \int_{-\infty}^{+\infty} \frac{\Omega}{1 - e^{-\beta\hbar\Omega}} \frac{1}{\Omega} \frac{\lambda\triangle_\infty(\Omega)}{m\Omega + \lambda^2\triangle_\infty(\Omega)} \Big|_{\omega-i0}^{\omega+i0} d\omega. \quad (1.55)$$

The right-hand side of this equation does not depend on ε when $0 < \varepsilon < 2\pi/\hbar\beta$. Thus, assuming $\varepsilon > 0$, $\varepsilon \to 0$, we derive from (1.50)

$$F_{\text{int}} = -\frac{3i\hbar}{2\pi} \int_0^1 d\lambda \int_{-\infty}^{+\infty} \frac{\Omega}{1 - e^{-\beta\hbar\Omega}} \frac{1}{\Omega} \frac{\lambda\triangle_\infty(\Omega)}{m\Omega + \lambda^2\triangle_\infty(\Omega)} \Big|_{\omega-i0}^{\omega+i0} d\omega. \quad (1.56)$$

Consider now the special "single-frequency" case

$$E(\omega) = \frac{K_0^2}{2} \{\delta(\omega - \nu_0) + \delta(\omega + \nu_0)\}. \tag{1.57}$$

Then

$$\triangle_\infty(\Omega) = -\frac{K_0^2 \Omega}{\Omega^2 - \nu_0^2},$$

and

$$-\frac{1}{\Omega} \frac{\lambda \triangle_\infty(\Omega)}{m\Omega + \lambda^2 \triangle_\infty(\Omega)} = \frac{1}{\Omega} \frac{\lambda K_0^2}{m\Omega^2 - (m\nu_0^2 + \lambda^2 K_0^2)}.$$

This expression has three poles:

$$\Omega = 0, \quad \Omega = \mu(\lambda), \quad \Omega = -\mu(\lambda),$$

where

$$\mu(\lambda) = \left(\nu_0^2 + \lambda^2 \frac{K_0^2}{m}\right)^{1/2}.$$

Let us notice that

$$\frac{1}{\Omega^2 - \mu^2(\lambda)} = \frac{1}{2\mu(\lambda)} \left(\frac{1}{\Omega - \mu(\lambda)} - \frac{1}{\Omega + \mu(\lambda)}\right).$$

Because of the obvious identity

$$-\frac{1}{\Omega} \frac{\lambda \triangle_\infty(\Omega)}{m\Omega + \lambda^2 \triangle_\infty(\Omega)} \Bigg|_{\omega - i0}^{\omega + i0}$$

$$= \frac{2\pi i \delta(\omega) \lambda K_0^2}{m\nu_0^2 + \lambda^2 K_0^2} - \frac{2\pi i \lambda K_0^2}{2\mu^2(\lambda)m} \{\delta(\omega - \mu(\lambda)) + \delta(\omega + \mu(\lambda))\}$$

and

$$-\frac{3i\hbar}{2\pi} \int_{-\infty}^{+\infty} \frac{\Omega}{1 - e^{-\beta\hbar\Omega}} \frac{1}{\Omega} \frac{\lambda \triangle_\infty(\Omega)}{m\Omega + \lambda^2 \triangle_\infty(\Omega)} \Bigg|_{\omega - i0}^{\omega + i0} d\omega$$

$$= -3\vartheta \frac{\lambda K_0^2}{m\nu_0^2 + \lambda^2 K_0^2} + \frac{3\lambda K_0^2 \hbar}{2m\mu(\lambda)} \left(\frac{1}{1 - e^{-\beta\hbar\mu(\lambda)}} + \frac{e^{-\beta\hbar\mu(\lambda)}}{1 - e^{-\beta\hbar\mu(\lambda)}}\right)$$

$$= \frac{d}{d\lambda} \left(-\frac{3\vartheta}{2} \ln(m\nu_0^2 + \lambda^2 K_0^2) + \frac{3}{2}\vartheta \ln(1 - e^{-\beta\hbar\mu(\lambda)}) + \frac{3}{2}\vartheta \ln(e^{\beta\hbar\mu(\lambda)} - 1)\right).$$

Therefore [f]

$$F_{\text{int}} = -3\vartheta \ln \left(\frac{m + K_0^2/\nu_0^2}{m} \right)^{1/2} + \frac{3\hbar}{2} (\mu - \nu_0) - 3\vartheta \ln \frac{1 - e^{-\beta\hbar\nu_0}}{1 - e^{-\beta\hbar\mu}}, \quad (1.58)$$

where

$$\mu = \left(\nu_0^2 + \frac{K_0^2}{m} \right)^{1/2} \equiv \mu(\lambda = 1).$$

As we have seen by now, all considered Green functions and correlation functions relating to the particle S, as well as the free energy F_{int}, are determined by one and the same function $E(\nu)$. The influence of the phonon field upon these quantities depends exclusively on this spectral intensity $E(\nu)$.

Thus, if we had two different systems of oscillators interacting with the given particle S (in the manner considered), for which the function $E(\nu)$ was one and the same, then all relevant quantities, mentioned above, would remain unchanged.

To illustrate this statement, it is worth considering a two-body problem, for example

$$H = \frac{\mathbf{p}^2}{2m} + \frac{K_0^2}{2} (\mathbf{r} - \mathbf{R})^2 + \frac{\mathbf{P}^2}{2M} = \sum_{\alpha=1}^{3} H_\alpha, \quad (1.59)$$

where

$$H_\alpha = \frac{p_\alpha^2}{2m} + \frac{K_0^2}{2} (r_\alpha - R_\alpha)^2 + \frac{P_\alpha^2}{2M},$$

and the corresponding "one-body" free Hamiltonians are:

$$H_S = \frac{\mathbf{p}^2}{2m} = \frac{1}{2m} \sum_{\alpha=1}^{3} p_\alpha^2,$$

$$H_\Sigma = \frac{\mathbf{P}^2}{2M} + \frac{K_0^2}{2} \mathbf{R}^2 = \sum_{\alpha=1}^{3} \left(\frac{P_\alpha^2}{2M} + \frac{K_0^2}{2} R_\alpha^2 \right).$$

---

[f] Here we have used the fact that

$$\frac{3}{2} \vartheta \int_0^1 d\lambda \frac{d}{d\lambda} \left\{ \ln \left(1 - e^{-\beta\hbar\mu(\lambda)} \right) + \ln \left(e^{\beta\hbar\mu(\lambda)} - 1 \right) \right\}$$

$$= \frac{3}{2} \vartheta \left(\ln \frac{e^{\beta\hbar\mu} - 1}{e^{\beta\hbar\nu_0} - 1} + \ln \frac{e^{\beta\hbar\mu} - 1}{e^{\beta\hbar\nu_0} - 1} \right)$$

$$= \frac{3}{2} \vartheta \left(\ln \frac{1 - e^{-\beta\hbar\mu}}{1 - e^{-\beta\hbar\nu_0}} + \ln \frac{e^{\beta\hbar\mu} e^{-\beta\hbar\nu_0} (1 - e^{-\beta\hbar\mu})}{1 - e^{-\beta\hbar\nu_0}} \right)$$

$$= -3\vartheta \ln \frac{1 - e^{-\beta\hbar\nu_0}}{1 - e^{-\beta\hbar\mu}} + \frac{3\hbar}{2} (\mu - \nu_0).$$

We shall calculate explicitly the free energy

$$F_{\text{int}} = -\vartheta \ln \frac{\text{Tr } e^{-\beta H}}{\text{Tr}_{(S)}\, e^{-\beta H_S}\, \text{Tr}_{(\Sigma)}\, e^{-\beta H_\Sigma}} = -3\vartheta \ln \frac{\text{Tr } e^{-\beta \mathcal{H}}}{\text{Tr}_{(S)}\, e^{-\beta \mathcal{H}_S}\, \text{Tr}_{(\Sigma)}\, e^{-\beta \mathcal{H}_\Sigma}},$$

where the one-dimensional Hamiltonians $\mathcal{H}$, $\mathcal{H}_S$ and $\mathcal{H}_\Sigma$ are given by the following expressions:

$$\mathcal{H} = \frac{p^2}{2m} + \frac{K_0^2}{2}(x - X)^2 + \frac{1}{2M}\, P^2,$$

$$\mathcal{H}_S = \frac{p^2}{2m}, \qquad \mathcal{H}_\Sigma = \frac{P^2}{2M} + \frac{K_0^2}{2}\, X^2.$$

To diagonalize the one-dimensional Hamiltonian $\mathcal{H}$, we introduce normal coordinates q, Q and corresponding normal momentum variables y, Y:

$$q = \frac{mx + MX}{m + M}, \quad Q = x - X. \tag{1.58a}$$

Noting that

$$\frac{\partial}{\partial x} = \frac{m}{m + M}\frac{\partial}{\partial q} + \frac{\partial}{\partial Q},$$

$$\tag{1.58b}$$

$$\frac{\partial}{\partial X} = \frac{M}{m + M}\frac{\partial}{\partial q} - \frac{\partial}{\partial Q},$$

we put

$$p = \frac{m}{m + M}\, y + Y, \qquad P = \frac{M}{m + M}\, y - Y. \tag{1.58c}$$

Substituting these results into (1.57), we arrive at the canonically transformed Hamiltonian

$$H = \frac{1}{2(m + M)}\, y^2 + \frac{1}{2}\frac{M + m}{Mm}\, Y^2 + \frac{K_0^2}{2}\, Q^2 = H_{\text{in}} + H_{\text{osc}}, \tag{1.61}$$

where

$$H_{\text{in}} = \frac{y^2}{2(m + M)}, \qquad H_{\text{osc}} = \frac{1}{2}\frac{M + m}{Mm}\, Y^2 + \frac{K_0^2}{2}\, Q^2.$$

Thus

$$F_{\text{int}} = -3\vartheta \ln \frac{\text{Tr } e^{-\beta H_{\text{in}}}}{\text{Tr } e^{-\beta H_S}} - 3\vartheta \ln \frac{\text{Tr } e^{-\beta H_{\text{osc}}}}{\text{Tr } e^{-\beta H_\Sigma}}.$$

It is well known that the free energy of the oscillator H_Σ with the frequency $\nu_0 = \left(K_0^2/M\right)^{1/2}$ in the one-dimensional case is [g]

$$F_\Sigma = \frac{\hbar \nu_0}{2} - \vartheta \ln \frac{1}{1 - e^{-\beta \hbar \nu_0}},$$

and because the oscillator H_{osc} has frequency

$$\mu = \left(\frac{K_0^2 (M+m)}{Mm} \right)^{1/2} = \nu_0 \left(1 + \frac{M}{m} \right)^{1/2},$$

then F_{osc} is

$$F_{\text{osc}} = \frac{\hbar \mu}{2} - \vartheta \ln \frac{1}{1 - e^{-\beta \hbar \mu}}.$$

Therefore

$$F_{\text{int}} = -3\vartheta \ln \frac{\text{Tr}\, e^{-\beta H_{\text{in}}}}{\text{Tr}\, e^{-\beta H_S}} - 3\vartheta \ln \frac{1 - e^{-\beta \hbar \nu_0}}{1 - e^{-\beta \hbar \mu}} + \frac{3}{2} \hbar(\mu - \nu_0).$$

Because the position x — belongs to the interval $-L/2 < x < L/2$, the corresponding momentum variable p can take only discrete values

$$\frac{2\pi}{L} n\hbar, \quad n = 0, \pm 1, \pm 2, \ldots,$$

and

$$\frac{\text{Tr}\, e^{-\beta H_{\text{in}}}}{\text{Tr}\, e^{-\beta H_S}} = \frac{\displaystyle\sum_{(n)} \exp \left\{ -\left(\frac{2\pi}{L} n\hbar \right)^2 \frac{\beta}{2(M+m)} \right\}}{\displaystyle\sum_{(n)} \exp \left\{ -\left(\frac{2\pi}{L} n\hbar \right)^2 \frac{\beta}{2m} \right\}}$$

$$\xrightarrow[L \to \infty]{} \frac{\dfrac{1}{2\pi\hbar} \displaystyle\int_{-\infty}^{+\infty} \exp \left(-\dfrac{\beta p^2}{2(M+m)} \right) dp}{\dfrac{1}{2\pi\hbar} \displaystyle\int_{-\infty}^{+\infty} \exp \left(-\dfrac{\beta p^2}{2m} \right) dp} = \left(\frac{m+M}{m} \right)^{1/2}.$$

Finally, we obtain the expression for the free energy:

$$F_{\text{int}} = -3\vartheta \ln \left(\frac{m+M}{m} \right)^{1/2} - 3\vartheta \ln \frac{1 - e^{-\beta \hbar \nu_0}}{1 - e^{-\beta \hbar \mu}} + \frac{3}{2} \hbar(\mu - \nu_0).$$

---

[g] For the Hamiltonian

$$H = \frac{P^2}{2m} + \frac{m\omega^2 X^2}{2}$$

the free energy is given by the expression

$$F = \frac{\hbar\omega}{2} - \vartheta \ln \left(1 - e^{-\beta \hbar \omega} \right), \quad \vartheta = K_\beta T.$$

This expression coincides with (1.56) if one puts

$$\frac{K_0^2}{\nu_0^2} = M.$$

Observation. It is easy to obtain (1.34) for the correlator starting from the Hamiltonian of the two-body problem (1.57). Put

$$Y = i(a^\dagger - a)\left(\frac{\hbar \mathcal{M}\mu}{2}\right)^{1/2}, \qquad Q = (a^\dagger + a)\left(\frac{\hbar}{2\mathcal{M}\mu}\right)^{1/2}, \qquad (1.62)$$

where a, $a^\dagger$ are Bose-amplitudes,

$$\mathcal{M} = \frac{Mm}{M + m}.$$

and the mass parameter μ is determined by (1.59):

$$\mu^2 = \frac{K_0^2}{\mathcal{M}}.$$

Further, we proceed with the oscillator Hamiltonian rewritten in new terms:

$$H_{\mathrm{osc}} = \frac{\hbar \mu}{2} + \hbar \mu a^\dagger a,$$

and

$$i\hbar \frac{da}{dt} = \hbar \mu a, \qquad i\hbar \frac{da^\dagger}{dt} = -\hbar \mu a^\dagger,$$

$$a(t) = e^{-i\mu t} a, \qquad a^\dagger(t) = e^{i\mu t} a^\dagger,$$

since

$$\frac{dy}{dt} = 0, \quad y = \mathrm{const.}$$

It follows from (1.58a) that

$$p_\alpha(t) = \frac{m}{m + M} y + i(a^\dagger e^{i\mu t} - a e^{-i\mu t})\left(\frac{\hbar \mathcal{M}\mu}{2}\right)^{1/2}. \qquad (1.63)$$

From this,

$$\langle p_\alpha(t) p_\alpha(\tau) \rangle_{\mathrm{eq}} = \left(\frac{m}{m + M}\right)^2 \langle y^2 \rangle_{\mathrm{eq}} + \frac{\hbar \mathcal{M}\mu}{2}(\langle aa^\dagger \rangle e^{-i\mu(t-\tau)} + \langle a^\dagger a \rangle e^{i\mu(t-\tau)}).$$
$$(1.64)$$

But

$$\langle y^2 \rangle_{\mathrm{eq}} = (m + M)\vartheta,$$

$$\langle aa^\dagger \rangle = \frac{1}{1 - e^{-\beta \hbar \mu}}, \qquad \langle a^\dagger a \rangle = \frac{1}{e^{\beta \hbar \mu} - 1}, \qquad \mathcal{M}\mu = \frac{\mathcal{M}\mu^2}{\mu} = \frac{K_0^2}{\mu},$$

and we arrive at the same equation (1.34) as before.

Therefore we come to the conclusion that two-time equilibrium correlation functions of the particle S variables are the same for the Hamiltonian (1.1) in the single-frequency case то $\nu(f) = \nu_0$ and when

$$E(\omega) = \frac{K^2}{2} [\delta(\omega - \nu_0) + \delta(\omega + \omega_0)],$$

as in the case of the two-body problem Hamiltonian (1.57).

We now return to the expression for the free energy F_{int} in the single-frequency case and consider the passage to classical mechanics. By setting $\hbar \to 0$, in (1.56), we get the classical result for the free energy:

$$F_{\text{int}} = 0.$$

It is obvious that the part of the free energy that is due to interaction is always zero in classical mechanics for dynamical systems described by a Hamiltonian of the kind (1.1) in the case $K^2 = K_0^2 + \eta^2$. This statement can be proved starting from (1.53), which can be rewritten as

$$\left\langle \frac{\partial H(\lambda)}{\partial \lambda} \right\rangle_{\lambda,\text{eq}} = -\frac{3i\hbar}{2\pi} \int_{-\infty}^{+\infty} \frac{\Omega}{1 - e^{-\beta\hbar\Omega}} \frac{\lambda\triangle(\Omega)}{m\Omega^2 - \eta^2 + \lambda^2\Omega\triangle(\Omega)} \Big|_{\omega-i0}^{\omega+i0} d\omega, \tag{1.65}$$

where

$$0 < \varepsilon < \frac{2\pi}{\hbar\beta}.$$

In the classical limit,

$$\lim_{\hbar \to 0} \frac{\hbar\Omega}{1 - e^{-\beta\hbar\Omega}} = \frac{1}{\beta} = \vartheta.$$

Consequently, it is true for the classical mechanics that

$$\left\langle \frac{\partial H(\lambda)}{\partial \lambda} \right\rangle_{\lambda,\text{eq}} = -\frac{3i\vartheta}{2\pi} \left(\int_{i\varepsilon-\infty}^{i\varepsilon+\infty} F(\Omega)\, d\Omega - \int_{-i\varepsilon-\infty}^{-i\varepsilon+\infty} F(\Omega)\, d\Omega \right), \tag{1.66}$$

where

$$F(\Omega) = \frac{\lambda\triangle(\Omega)}{m\Omega^2 - \eta^2 + \lambda^2\Omega\triangle(\Omega)}. \tag{1.67}$$

It should be observed that $F(\Omega)$ is a regular analytic function on the half-plane

$$\text{Im}\,(\Omega) \geqslant \varepsilon > 0.$$

Thus

$$\int_{\mathcal{L}} F(\Omega)\, d\Omega = 0 \tag{1.68}$$

for any closed contour $\mathcal{L}$ lying in this half-plane. Let us take for $\mathcal{L}$ the contour composed of the interval $(i\varepsilon - L, i\varepsilon + L)$ and the half-circle C

Fig. 1.2.   Fig. 1.3.

with center at the point $i\varepsilon$ and of radius L (see Fig. 1.2). On this contour,

$$F(\Omega) = \mathcal{O}\left(\frac{1}{L^2}\right),$$

so we can see that

$$\int_C F(\Omega)\,d\Omega = \mathcal{O}\left(\frac{1}{L}\right) \to 0.$$

Therefore

$$\int_{i\varepsilon-\infty}^{i\varepsilon+\infty} F(\Omega)\,d\Omega = 0.$$

The same considerations may be applied to prove that

$$\int_{-i\varepsilon-\infty}^{-i\varepsilon+\infty} F(\Omega)\,d\Omega.$$

In fact, $F(\Omega)$ — is a regular analytic function on the lower half-plane

$$\operatorname{Im}\Omega \leqslant -\varepsilon < 0.$$

Therefore it is sufficient to choose the proper contour (see Fig. 1.3) and to repeat all the previous reasoning.

Thus, taking (1.50) and (1.64) into account, we have

$$F_{\text{int}} = 0.$$

It should be stressed that this result follows entirely from the treatment of the dynamical system within the framework of classical mechanics.

In the opposite, quantum mechanical case, (1.63) indicates that the function

$$F(\Omega) = \frac{\hbar\Omega}{1 - e^{-\beta\hbar\Omega}} \frac{\lambda\triangle(\Omega)}{m\Omega^2 - \eta^2 + \lambda^2\Omega\triangle(\Omega)}$$

has an infinite number of poles on the imaginary axis:

$$\Omega = \frac{2\pi i n}{\hbar\beta}, \quad \text{for } n \text{ integer}, \tag{1.69}$$

and for this reason integrals of the type (1.66) are not equal to zero (they are equal to the sum of residues taken at the poles (1.67)).

1.5. Average Values of T-products

Consider now equilibrium averages of the operator products

$$\langle T\left\{[r_\alpha(t) - r_\alpha(\tau)][r_{\alpha'}(t) - r_{\alpha'}(\tau)]\right\}\rangle_{eq},$$

where T denotes the "T-product" (i.e. the product of operators ordered in time). By definition,

$$T\{A(t_1)B(t_2)\} = \begin{cases} A(t_1)B(t_2), & t_1 > t_2, \\ B(t_2)A(t_1), & t_2 > t_1. \end{cases} \tag{1.70}$$

Because

$$[r_\alpha(t) - r_\alpha(\tau)][r_{\alpha'}(t) - r_{\alpha'}(\tau)]$$
$$= r_\alpha(t)r_{\alpha'}(t) - r_\alpha(\tau)r_{\alpha'}(t) - r_\alpha(t)r_{\alpha'}(\tau) + r_\alpha(\tau)r_{\alpha'}(\tau),$$

the following relation holds:

$$T\left\{[r_\alpha(t) - r_\alpha(\tau)][r_{\alpha'}(t) - r_{\alpha'}(\tau)]\right\}$$
$$= \begin{cases} r_\alpha(t)r_{\alpha'}(t) - r_{\alpha'}(t)r_\alpha(\tau) - r_\alpha(t)r_{\alpha'}(\tau) + r_\alpha(\tau)r_{\alpha'}(\tau), & t > \tau, \\ r_\alpha(t)r_{\alpha'}(t) - r_\alpha(\tau)r_{\alpha'}(t) - r_{\alpha'}(\tau)r_\alpha(t) + r_\alpha(\tau)r_{\alpha'}(\tau), & t < \tau. \end{cases}$$

Therefore, putting $\lambda = 1$ in (1.52), we get

$$\langle T\left\{[r_\alpha(t) - r_\alpha(\tau)][r_{\alpha'}(t) - r_{\alpha'}(\tau)]\right\}\rangle_{eq} = 0 \quad \text{if} \quad \alpha \neq \alpha'$$

and

$$\langle T\{[r_\alpha(t) - r_\alpha(\tau)]^2\}\rangle_{eq}$$
$$= \frac{i\hbar}{2\pi} \int_{-\infty}^{+\infty} \frac{2(1 - e^{-i\omega(t-\tau)})}{1 - e^{-\beta\hbar\omega}} \frac{1}{m\Omega^2 - \eta^2 + \Omega\Delta(\Omega)} \Big|_{\omega-i0}^{\omega+i0} d\omega \quad \text{if} \quad t > \tau,$$

$$\langle T\{[r_\alpha(t) - r_\alpha(\tau)]^2\}\rangle_{eq}$$
$$= \frac{i\hbar}{2\pi} \int_{-\infty}^{+\infty} \frac{2(1 - e^{-i\omega(\tau-t)})}{1 - e^{-\beta\hbar\omega}} \frac{1}{m\Omega^2 - \eta^2 + \Omega\Delta(\Omega)} \Big|_{\omega-i0}^{\omega+i0} d\omega \quad \text{if} \quad t < \tau.$$

Thus

$$\langle T\{[r_\alpha(t) - r_\alpha(\tau)]^2\}\rangle_{eq}$$
$$= \frac{i\hbar}{2\pi} \int_{-\infty}^{+\infty} \frac{2(1 - e^{-i\omega|t-\tau|})}{1 - e^{-\beta\hbar\omega}} \frac{1}{m\Omega^2 - \eta^2 + \Omega\Delta(\Omega)} \Big|_{\omega-i0}^{\omega+i0} d\omega. \tag{1.71}$$

For some applications it is helpful to be able to calculate the ordered products of operators depending on an "imaginary-time" argument.

Setting $t = -is$ and choosing real s to be the ordering parameter, we define the T-product as

$$T\{r_\alpha(-is)r_{\alpha'}(-i\sigma)\} = \begin{cases} r_\alpha(-is)r_{\alpha'}(-i\sigma) & \text{if } s > \sigma, \\ r_{\alpha'}(-i\sigma)r_\alpha(-is) & \text{if } \sigma > s. \end{cases}$$

Taking (1.52) into account under the condition $\lambda = 1$, one obtains

$$\langle T\{[r_\alpha(-is) - r_\alpha(-i\sigma)][r_{\alpha'}(-is) - r_{\alpha'}(-i\sigma)]\}\rangle_{\text{eq}} = 0 \qquad (1.72)$$

if $\alpha \neq \alpha'$, and, in the opposite case,

$$\langle T\{[r_\alpha(-is) - r_\alpha(-i\sigma)]^2\}\rangle_{\text{eq}}$$
$$= \frac{i\hbar}{2\pi} \int_{-\infty}^{+\infty} \frac{2(1 - e^{-\omega|s-\sigma|})}{1 - e^{-\beta\hbar\omega}} \frac{1}{m\Omega^2 - \eta^2 + \Omega\triangle(\Omega)} \bigg|_{w-i0}^{w+i0} d\omega. \qquad (1.73)$$

Consider the single-frequency case:

$$E(\omega) = \frac{K_0^2}{2}\{\delta(\omega - \nu_0) + \delta(\omega + \nu_0)\}.$$

Here we are not allowed to use (1.69) and (1.71) directly, observing from the very start that the final result would be just the same as in the case of the two-body model (1.57). Rigorously speaking, we mean that

$$\langle r_\alpha(t)r_\alpha(\tau)\rangle_{\text{eq}} = \langle x(t)x(\tau)\rangle_{\text{eq}}.$$

It follows from (1.58) that

$$x = q + \frac{M}{m + M}Q.$$

Since the time evolution of $q(t)$ and $Q(t)$ is generated by independent Hamiltonians H_{in} and H_{osc} respectively (see (1.59)), we have the equality

$$\langle x(t)x(\tau)\rangle_{\text{eq}} = \langle q(t)q(\tau)\rangle_{\text{eq}} + \frac{M^2}{(m + M)^2}\langle Q(t)Q(\tau)\rangle_{\text{eq}}.$$

From which

$$\langle T\{[r_\alpha(-is) - r_\alpha(-i\sigma)]^2\}\rangle_{\text{eq}} = \langle T\{[q(-is) - q(-i\sigma)]^2\}\rangle_{\text{eq}}$$
$$+ \frac{M^2}{(m + M)^2}\langle T\{(Q(-is) - Q(-i\sigma))^2\}\rangle_{\text{eq}} \qquad (1.74)$$

Thanks to (1.59),

$$q(t) - q(\tau) = \frac{(t - \tau)}{(M + m)}y, \quad y = \text{const},$$
$$\langle y^2\rangle_{\text{eq}} = (M + m)\vartheta.$$

Hence

$$\langle [q(t) - q(\tau)]^2\rangle_{\text{eq}} = \frac{(t - \tau)^2\vartheta}{M + m}. \qquad (1.75)$$

We also have

$$q(t)q(\tau) - q(\tau)q(t) = \int_{\tau}^{t}\{q'(t)q(\tau) - q(\tau)q'(t)\}\,dt$$

$$= \frac{1}{M+m}\int_{\tau}^{t}\{yq(\tau) - q(\tau)y\}\,dt = \frac{t-\tau}{M+m}\,(yq - qy) = -i\hbar\frac{t-\tau}{M+m}. \quad (1.76)$$

Consequently,

$$\langle T\{[q(-is) - q(-i\sigma)]^2\}\rangle_{\text{eq}} = \langle q^2(-is) - 2q(-is)q(-i\sigma) + q^2(-i\sigma)\rangle_{\text{eq}}$$

$$= \langle[q(-is) - q(-i\sigma)]^2\rangle_{\text{eq}}$$

$$+ \langle -q(-is)q(-i\sigma) + q(-i\sigma)q(-is)\rangle_{\text{eq}} \quad \text{if} \quad s > \sigma,$$

and, thanks to (1.73) and (1.74),

$$\langle T\{[q(-is) - q(-i\sigma)]^2\}\rangle_{\text{eq}}$$

$$= -\frac{(s-\sigma)^2}{M+m}\vartheta + \frac{\hbar}{M+m}\,(s-\sigma) \quad \text{for} \quad s > \sigma.$$

Making the permutation $s \underset{\leftarrow}{\to} \sigma$, it is easy to show that

$$\langle T\{[q(-is) - q(-i\sigma)]^2\}\rangle_{\text{eq}} = -\frac{(s-\sigma)^2}{M+m}\vartheta + \frac{\hbar}{M+m}|s-\sigma|. \quad (1.77)$$

We must now find an explicit expression for the ordered correlator

$$\langle T\{[Q(-is) - Q(-i\sigma)]^2\}\rangle_{\text{eq}}.$$

in order to calculate the left-hand side of (1.72). Let us note that

$$Q(t) = (e^{i\mu t}a^{\dagger} + e^{-i\mu t}a)\left(\frac{\hbar}{2\mathcal{M}\mu}\right)^{1/2},$$

from which it follows that

$$\langle T\{[Q(-is) - Q(-i\sigma)]^2\}\rangle_{\text{eq}}$$

$$= \frac{\hbar}{2\mathcal{M}\mu}\{(e^{\mu s}a^{\dagger} + e^{-\mu s}a)(e^{\mu s}a^{\dagger} + e^{-\mu s}a)$$

$$+ (e^{\mu\sigma}a^{\dagger} + e^{-\mu\sigma}a)(e^{\mu\sigma}a^{\dagger} + e^{-\mu\sigma}a) - 2T(e^{\mu s}a^{\dagger} + e^{-\mu s}a)(e^{\mu\sigma}a^{\dagger} + e^{-\mu\sigma}a)\}.$$

Hence

$$\langle T\{[Q(-is) - Q(-i\sigma)]^2\}\rangle_{\text{eq}}$$

$$= \frac{\hbar}{2\mathcal{M}\mu}\,(2\langle a^{\dagger}a\rangle + 2\langle aa^{\dagger}\rangle - 2e^{\mu|s-\sigma|}\langle a^{\dagger}a\rangle - 2e^{-\mu|s-\sigma|}\langle aa^{\dagger}\rangle). \quad (1.78)$$

Here

$$\langle a^{\dagger}a\rangle = \frac{1}{e^{\beta\hbar\mu} - 1}, \quad \langle aa^{\dagger}\rangle = 1 + \frac{1}{e^{\beta\hbar\mu} - 1}.$$

Since

$$\frac{1}{m+M} = \frac{1}{Mm}\mathcal{M} = \frac{K_0^2}{\mu^2 Mm} = \left(\frac{\nu_0}{\mu}\right)^2\frac{1}{m}$$

(recalling that $\mathcal{M} = mM/(m+M)$) and

$$\frac{M^2}{(m+M)^2}\frac{1}{\mathcal{M}\mu} = \frac{\mathcal{M}^2}{m^2}\frac{1}{\mathcal{M}\mu} = \frac{\mathcal{M}}{m^2\mu} = \frac{K_0^2}{m^2\mu^3} = \frac{\mu^2 - \nu_0^2}{m\mu^3},$$

we find from (1.72), (1.75) and (1.76) that

$$\langle T\{[r_\alpha(-is) - r_\alpha(-i\sigma)]^2\}\rangle_{eq}$$

$$= -\left(\frac{\nu_0}{\mu}\right)^2\frac{\vartheta}{m}(s-\sigma)^2 + \frac{\hbar}{m}\left(\frac{\nu_0}{m}\right)^2|s-\sigma|$$

$$+ \frac{\mu^2-\nu_0^2}{m\mu^3}\hbar\left(\frac{1}{1-e^{-\beta\hbar\mu}}(1-e^{-\mu|s-\sigma|}) - \frac{1}{e^{\beta\hbar\mu}-1}(e^{\mu|s-\sigma|}-1)\right).$$

$$(1.79)$$

It is interesting to note that

$$\langle T\{[r_\alpha(-is) - r_\alpha(-i\sigma)]^2\}\rangle_{eq} \geqslant 0 \quad \text{if} \quad |s-\sigma| \leqslant \beta\hbar = \hbar/\vartheta, \qquad (1.80)$$

where $\beta = 1/\vartheta$ and $\vartheta = K_\beta T$.

1.6. Averaged Operator T-Product Calculus for Some Model Oscillatory Systems

Let the Hamiltonian Γ be a quadratic positive-definite form composed of Bose operators b_α, $b_\alpha^\dagger$. We denote the statistical sum as

$$Z = \text{Tr } e^{-\beta\Gamma}$$

and consider linear forms composed of the Bose operators b_α, $b_\alpha^\dagger$:

$$A_1, A_2, \ldots, A_s,$$

and statistical averages consisting of the products of these linear forms:

$$\langle A_1 A_2 \cdots A_3\rangle_\Gamma = Z^{-1}e^{-\beta\Gamma}(A_1 A_2 \cdots A_s). \qquad (1.81)$$

Let us apply to (1.79) the well-known Bloch–Dominicis theorem, which generalizes the Wick theorem. If we introduce couplings of the type

$$\overline{A_j A_l} = \langle A_j A_l\rangle_\Gamma,$$

we see that the expression (1.79) is equal to the sum of products of all possible couplings. For example,

$$\langle A_1 A_2 A_3 A_4\rangle_\Gamma = \langle \overline{A_1 \overline{A_2 A_3} A_4}\rangle_\Gamma + \langle \overline{A_1 A_2 \overline{A_3 A_4}}\rangle_\Gamma + \langle \overline{A_1 \overline{A_2 A_3} A_4}\rangle_\Gamma$$

$$= \langle A_1 A_2\rangle_\Gamma\langle A_3 A_4\rangle_\Gamma + \langle A_1 A_3\rangle_\Gamma\langle A_2 A_4\rangle_\Gamma + \langle A_1 A_4\rangle_\Gamma\langle A_2 A_3\rangle_\Gamma. \qquad (1.82)$$

Of course, the expression (1.79) is zero if s is odd, because in this case one of the operators $A_1, A_2, ..., A_s$ is left uncoupled, and

$$\langle A_j \rangle_\Gamma = 0,$$

because A_j is a linear form composed of operators b_α and $b_\alpha^\dagger$ and Γ is a quadratic form.

Then we apply this well-known technique to the calculation of the expression

$$\langle e^A \rangle_\Gamma,$$

where A is some linear form composed of the above-mentioned Bose operators. We arrive at the following result:

$$\langle e^A \rangle_\Gamma = \sum_{n=0}^\infty \frac{1}{n!} \langle A^n \rangle_\Gamma = \sum_{k=0}^\infty \frac{1}{(2k)!} \langle A^{2k} \rangle_\Gamma = 1 + \sum_{k=1}^\infty \frac{1}{(2k)!} \langle A^{2k} \rangle_\Gamma. \quad (1.83)$$

Thanks to the Bloch–Dominicis theorem,

$$\langle A^{2k} \rangle_\Gamma = G(k) \langle A^2 \rangle_\Gamma^k,$$

where $G(k)$ is the number of all possible couplings in the expression

$$\langle A_1 \cdots A_s \rangle_\Gamma.$$

One can see that

$$G(1) = 1, \quad G(2) = 3, \quad G(k+1) = (2k+1)G(k).$$

Thus

$$G(k) = 1 \cdot 3 \cdots (2k-1), \quad \frac{G(k)}{(2k!)} = \frac{1 \cdot 3 \cdots (2k-1)}{1 \cdot 2 \cdot 3 \cdot 4 \cdots (2k)} = \frac{1}{2^k k!}.$$

From where

$$\langle e^A \rangle_\Gamma = 1 + \sum_{k=1}^\infty \frac{1}{k!} \langle \frac{1}{2} A^2 \rangle_\Gamma^2 = e^{\frac{1}{2} \langle A^2 \rangle_\Gamma}. \quad (1.84)$$

We are now going to consider T-products of operators ordered in the parameter s. By definition,

$$T\{A(s_1)A(s_2)\} = \begin{cases} A(s_1)A(s_2) & \text{if } s_1 > s_2, \\ A(s_2)A(s_1) & \text{if } s_2 > s_1, \end{cases}$$

and respectively

$$TA(s_1)A(s_2) \cdots A(s_n) = A(s'_1) \cdots A(s'_n),$$

where $s'_1, ..., s'_n$ is just the same set of parameters $s_1, ..., s_n$, but ordered in time in the following way:

$$s'_1 \geqslant s'_2 \geqslant ... \geqslant s'_n.$$

It is interesting to note that, thanks to the definition of the T-product, the operators $A(s_j)$ commute under the sign of the T-product. For example,

$$T\{A(s_1)A(s_2)\} = T\{A(s_2)A(s_1)\}.$$

Now we should try to simplify an expression of the kind

$$\left\langle T\{e^{\int_{s_0}^{s_1} ds\, A(s)}\}\right\rangle_\Gamma, \quad s_0 < s_1,$$

where the averaging is with respect to the quadratic in Bose operators Hamiltonian Γ, (we have in mind here the same Hamiltonian as that one presented in (1.79)) and $A(s)$ are linear forms composed of these Bose operators with coefficients dependent on the ordering parameter s.

$$\left\langle T\left\{e^{\int_{s_0}^{s_1} ds\, A(s)}\right\}\right\rangle_\Gamma = \sum_{n=0}^\infty \left\langle T\left\{\left(\int_{s_0}^{s_1} ds\, A(s)\right)^n\right\}\right\rangle_\Gamma.$$

Keeping in mind that the Bloch–Dominicis theorem can be applied not only to the ordinary products but also to the T-products, we may repeat our previous reasoning and write down the final result at once:

$$\left\langle T\left\{e^{\int_{s_0}^{s_1} ds\, A(s)}\right\}\right\rangle_\Gamma = \exp\left(\frac{1}{2}\left\langle T\left\{\left(\int_{s_0}^{s_1} ds\, A(s)\right)^2\right\}\right\rangle_\Gamma\right).$$

From the other side,

$$T\left\{\left(\int_{s_0}^{s_1} A(s)\right)^2\right\} = T\left\{\int_{s_0}^{s_1} ds \int_{s_0}^{s_1} d\sigma\, A(s)A(\sigma)\right\} = \int_{s_0}^{s_1} ds \int_{s_0}^{s_1} d\sigma\, T\{A(s)A(\sigma)\}.$$

Therefore

$$\left\langle T\left\{e^{\int_{s_0}^{s_1} ds\, A(s)}\right\}\right\rangle_\Gamma = \exp\left(\frac{1}{2}\int_{s_0}^{s_1} ds \int_{s_0}^{s_1} d\sigma\, \langle T\{A(s)A(\sigma)\}\rangle_\Gamma\right). \tag{1.85}$$

We can now apply these results to the case of the oscillator Hamiltonian

$$\Gamma = \frac{p^2}{2m} + \frac{m\omega^2}{2} q^2. \tag{1.86}$$

Put

$$Q(s) = e^{\frac{s}{\hbar}\Gamma} q e^{-\frac{s}{\hbar}\Gamma}, \tag{1.87}$$

and let $\lambda(s)$ be a c — number function dependent on s. We want to calculate the expression

$$\left\langle T\left\{e^{\int_0^{\beta\hbar} \lambda(s)Q(s)\, ds}\right\}\right\rangle_\Gamma \tag{1.88}$$

Note that the Heisenberg equation for the time-dependent variable $q(t)$ is

$$i\hbar \frac{dq(t)}{dt} = q(t)\Gamma - \Gamma q(t), \quad q(0) = q,$$

from which it follows that

$$q(t) = e^{\frac{it}{\hbar}\Gamma} q e^{-\frac{it}{\hbar}\Gamma}.$$

Equation (1.84) states that

$$Q(s) = q(-is), \tag{1.89}$$

so that s could be treated as an "imaginary time". Introducing Bose amplitudes, we have

$$q = \left(\frac{\hbar}{2m\omega}\right)^{1/2} (b + b^\dagger), \qquad p = i \left(\frac{2m\omega}{\hbar}\right)^{1/2} (b^\dagger - b) \tag{1.90}$$

and

$$\Gamma = \hbar\omega b^\dagger b + \frac{\hbar\omega}{2}.$$

Here Γ is represented by the sum of the quadratic form $\hbar\omega b^\dagger b$ and the constant term $\hbar\omega/2$. It is clear that the constant term does not influence the calculation of averages of the type:

$$\langle \ldots \rangle_\Gamma = \frac{\mathrm{Tr}(\ldots e^{-\beta\Gamma})}{\mathrm{Tr}\, e^{-\beta\Gamma}} = \langle \ldots \rangle_{\hbar\omega b^\dagger b}.$$

Therefore we can apply (1.82) to calculate the expression (1.85):

$$\left\langle T\left\{ e^{\int_0^{\beta\hbar} \lambda(s)Q(s)\,ds} \right\} \right\rangle_\Gamma = \exp\left(\frac{1}{2} \int_0^{\hbar\beta} ds \int_0^{\hbar\beta} d\sigma\, \lambda(s)\lambda(\sigma)\langle T\{Q(s)Q(\sigma)\}\rangle_\Gamma \right). \tag{1.91}$$

Noting that

$$b(t) = e^{-i\omega t} b, \qquad b^\dagger(t) = e^{i\omega t} b^\dagger,$$

we obtain from (1.86) and (1.87) that

$$Q(s) = \left(\frac{\hbar}{2m\omega}\right)^{1/2} (e^{-\omega s} b + e^{\omega s} b^\dagger)$$

and

$$\langle T\{Q(s)Q(\sigma)\}\rangle_\Gamma = \frac{\hbar}{2m\omega} (e^{-\omega|s-\sigma|}\langle bb^\dagger \rangle_\Gamma + e^{\omega|s-\sigma|}\langle b^\dagger b \rangle_\Gamma)$$

$$= \frac{\hbar}{2m\omega} (1 - e^{-\beta\omega\hbar})^{-1}(e^{-\omega|s-\sigma|} + e^{-\omega\beta\hbar+\omega|s-\sigma|}). \tag{1.92}$$

It follows from here and from (1.88) that

$$\left\langle T\left\{ e^{\int_0^{\beta\hbar} \lambda(s)Q(s)\,ds} \right\} \right\rangle_\Gamma$$

$$= \exp\left(\int_0^{\hbar\beta} ds \int_0^{\hbar\beta} d\sigma \frac{\hbar\lambda(s)\lambda(\sigma)(e^{-\omega|s-\sigma|} + e^{-\omega\beta\hbar+\omega|s-\sigma|})}{4m\omega(1 - e^{-\beta\omega\hbar})} \right). \tag{1.93}$$

Consider the expression

$$K(s, \sigma) = e^{-\omega|s-\sigma|} + e^{-\omega\beta\hbar+\omega|s-\sigma|} \tag{1.94}$$

for

$$0 < s < \beta\hbar, \quad 0 < \sigma < \beta\hbar \tag{1.95}$$

as a function of s. We get

$$K(0, \sigma) = e^{-\omega\sigma} + e^{-\omega\beta\hbar+\omega\sigma},$$

$$K(\beta\hbar, \sigma) = e^{-\omega(\beta\hbar-\sigma)} + e^{-\omega\beta\hbar+\omega\beta\hbar-\omega\sigma},$$

whence

$$K(0, \sigma) = K(\beta\hbar, \sigma). \tag{1.96}$$

Since

$$\frac{d}{ds}(s - \sigma) = 1, \quad \frac{d}{ds}(\sigma - s) = -1,$$

we see that

$$\frac{d}{ds}|s - \sigma| = \varepsilon(s - \sigma),$$

where

$$\varepsilon(s - \sigma) = \begin{cases} 1, & s > \sigma \\ -1, & s < \sigma. \end{cases}$$

Therefore differentiation of the expression (1.91) yields

$$\frac{d}{ds}K(s, \sigma) = \varepsilon(s - \sigma)(-\omega e^{-\omega|s-\sigma|} + \omega e^{-\omega\beta\hbar+\omega|s-\sigma|}).$$

Keeping in mind the trivial relations

$$\varepsilon^2(s - \sigma) = 1, \quad \frac{d\varepsilon(s - \sigma)}{ds} = 2\delta(s - \sigma),$$

where $\delta(s - \sigma)$ — is the usual Dirac delta function, and taking (1.91) into account, we obtain

$$\frac{d^2 K(s, \sigma)}{ds^2} - \omega^2 K(s, \sigma) = -2\omega(1 - e^{-\beta\omega\hbar})\delta(s - \sigma). \tag{1.97}$$

From here, for any σ, belonging to the interval $(0, \beta\hbar)$, the function $K(s, \sigma)$, taken as a function of s, satisfies the differential equation

$$\frac{d^2 y(s)}{ds^2} - \omega^2 y(s) = -2\omega(1 - e^{-\beta\omega\hbar})\delta(s - \sigma) \tag{1.98}$$

with boundary conditions

$$y(0) = y(\beta\hbar), \quad y'_s(0) = y'_s(\beta\hbar). \tag{1.99}$$

The first condition follows from (1.93), while the second is a consequence of the identities

$$K'_s(0, \sigma) = e^{-\omega\sigma} + e^{-\omega\beta\hbar + \omega\sigma},$$
$$K'_s(\beta\hbar, \sigma) = -\omega e^{-\omega(\beta\hbar - \sigma)} + \omega e^{-\omega\beta\hbar + \omega(\beta\hbar - \sigma)} = K'_s(0, \sigma), \qquad (1.100)$$
$$0 < \sigma < \beta\hbar.$$

It is easy to see that equation (1.95) with boundary conditions (1.96) cannot have two different solutions. Suppose, on the contrary, that we could find two different solutions $y_1(s)$, $y_2(s)$. Taking their difference

$$y_1(s) - y_2(s) = Z(s),$$

we construct a nontrivial solution $Z(s)$ that satisfies the equation

$$\frac{d^2 Z(s)}{ds^2} - \omega^2 Z(s) = 0, \qquad (1.101)$$

with boundary conditions

$$Z(0) = Z(\beta\hbar), \qquad Z'_s(0) = Z'_s(\beta\hbar). \qquad (1.102)$$

However, it follows from (1.98) that

$$Z(s) = Ae^{-\omega s} - Be^{\omega s},$$

whereas, from (1.99), we must have

$$A(1 - e^{-\beta\omega\hbar}) + B(e^{\omega\beta\hbar} - 1) = 0,$$
$$\omega A(1 - e^{-\beta\omega\hbar}) + \omega B(e^{\omega\beta\hbar} - 1) = 0. \qquad (1.103)$$

The determinant of these two linear uniform equations is nonzero:

$$\det = 2\omega(1 - e^{-\beta\omega\hbar})(e^{\omega\beta\hbar} - 1)$$

Thus (1.100) has only the trivial solution

$$A = 0, \quad B = 0,$$

and consequently

$$Z(s) = 0.$$

Thus we have proved that the differential equation (1.95) with boundary conditions (1.96) cannot have two different solutions. Remembering that $0 < \sigma < \beta\hbar$, we can rewrite (1.95) in the form

$$\frac{d^2 y(s)}{ds^2} - \omega^2 y(s) = -2\omega(1 - e^{-\beta\omega\hbar})\frac{1}{\beta\hbar}\sum_{(n)} e^{in\frac{s-\sigma}{\beta\hbar}2\pi}. \qquad (1.104)$$

To satisfy the boundary conditions, we must put

$$y(s) = \sum_{(n)} C_n e^{in\frac{s}{\beta\hbar}2\pi},$$

and, substituting this ansatz into (1.101), we arrive at

$$\left\{\left(\frac{2\pi n}{\beta\hbar}\right)^2 + \omega^2\right\} C_n = 2\omega(1 - e^{-\beta\omega\hbar})\frac{1}{\beta\hbar} e^{in\frac{\sigma}{\beta\hbar}2\pi}.$$

Since there are no other solutions of (1.95) with the boundary conditions (1.96), we see that

$$K(s,\sigma) = 2\omega\frac{1 - e^{-\beta\omega\hbar}}{\beta\hbar}\sum_{(n)}\frac{e^{in\frac{s-\sigma}{\beta\hbar}2\pi}}{(2\pi n/\beta\hbar)^2 + \omega^2}. \tag{1.105}$$

Thus, thanks to (1.90), we have

$$\left\langle T\left\{e^{\int_0^{\beta\hbar}\lambda(s)Q(s)\,ds}\right\}\right\rangle_\Gamma$$

$$= \exp\left(\frac{1}{m\beta}\int_0^{\hbar\beta}ds\int_0^{\hbar\beta}d\sigma\sum_{(n)}\frac{e^{in\frac{s-\sigma}{\beta\hbar}2\pi}}{(2\pi n/\beta\hbar)^2 + \omega^2}\lambda(s)\lambda(\sigma)\right)$$

$$= \exp\left(\frac{1}{m\beta}\sum_{(n)}\frac{\int_0^{\beta\hbar}e^{2\pi\frac{ins}{\beta\hbar}}\lambda(s)\,ds\int_0^{\beta\hbar}e^{-2\pi\frac{ins}{\beta\hbar}}\lambda(s)\,ds}{(2\pi n/\beta\hbar)^2 + \omega^2}\right). \tag{1.106}$$

Note that (1.102) was proved to be valid only for the domain

$$0 \leqslant s \leqslant \beta\hbar, \quad 0 < \sigma < \beta\hbar.$$

but, because the kernel $K(s,\sigma)$ is continuous and the Fourier series on the right-hand side of (1.102) converges absolutely and uniformly, we can prove that the expression (1.102) holds even for the closed domain

$$0 \leqslant s \leqslant \beta\hbar, \quad 0 \leqslant \sigma \leqslant \beta\hbar.$$

Let us now consider the case when, instead of the integral

$$\int_0^{\beta\hbar}\lambda(s)Q(s)\,ds$$

we have the finite sum

$$\sum_{1\leqslant j\leqslant N+1} i\nu_j Q(s_j),$$

where ν_j are real and

$$s_1 = 0, \quad s_{N+1} = \beta\hbar, \quad 0 < s_2 < s_3 < ... < s_N < \beta\hbar. \tag{1.107}$$

Repeating all of previous reasoning, we find

$$\left\langle T\left\{\exp\left(i\sum_{1\leqslant j\leqslant N+1} i\nu_j Q(s_j)\right)\right\}\right\rangle_\Gamma$$

$$= \exp\left(-\frac{1}{m\beta}\sum_{(n)}\frac{\left|\sum_{1\leqslant j\leqslant N+1} e^{\frac{2\pi i n s_j}{\beta\hbar}}\nu_j\right|^2}{(2\pi n/\beta\hbar)^2+\omega^2}\right). \quad (1.108)$$

Consider a quadratic form with respect to the variables ν_j:

$$\Omega(\ldots\nu_j\ldots) = \frac{1}{m\beta}\sum_{(n)}\frac{\left|\sum_{1\leqslant j\leqslant N+1} e^{\frac{2\pi i n s_j}{\beta\hbar}}\nu_j\right|^2}{(2\pi n/\beta\hbar)^2+\omega^2}. \quad (1.109)$$

Because of (1.104)

$$\sum_{1\leqslant j\leqslant N+1}\nu_j e^{\frac{2\pi i n s_j}{\beta\hbar}} = (\nu_1+\nu_{N+1}) + \sum_{2\leqslant j\leqslant N+1}\nu_j e^{\frac{2\pi i n s_j}{\beta\hbar}}. \quad (1.110)$$

Therefore

$$\Omega(\ldots\nu_j\ldots) = \frac{1}{m\beta}\sum_{(n)}\frac{\left|(\nu_1+\nu_{N+1}) + \sum_{2\leqslant j\leqslant N+1}\nu_j e^{\frac{2\pi i n s_j}{\beta\hbar}}\right|^2}{(2\pi n/\beta\hbar)^2+\omega^2}, \quad (1.111)$$

and hence

$$\Omega(\ldots\nu_j\ldots) \geqslant 0.$$

We shall show that

$$\Omega(\ldots\nu_j\ldots) = 0 \quad (1.112)$$

if and only if all the variables $\nu_1+\nu_{N+1}, \nu_2, \ldots, \nu_N$ are zero:

$$\nu_1+\nu_{N+1} = 0, \quad \nu_2 = 0, \quad \nu_3 = 0, \ldots, \nu_N = 0. \quad (1.113)$$

In fact, it follows directly from (1.108) that (1.109) holds only if, for any integer n (positive, negative or zero),

$$(\nu_1+\nu_{N+1}) + \sum_{2\leqslant j\leqslant N+1}\nu_j e^{\frac{2\pi i n s_j}{\beta\hbar}} = 0. \quad (1.114)$$

Let us sum these relations for each $n = 0, 1, \ldots, \Lambda$ and divide the result by $\Lambda + 1$. We have

$$\nu_1+\nu_{N+1} + \sum_{2\leqslant j\leqslant N+1}\frac{1}{\Lambda+1}\nu_j\frac{e^{\frac{2\pi i(\Lambda+1)s_j}{\beta\hbar}}-1}{e^{\frac{2\pi i s_j}{\beta\hbar}}-1} = 0.$$

Note that, thanks to (1.104),

$$e^{\frac{2\pi i s_j}{\beta\hbar}} - 1 \neq 0.$$

Hence, using the limiting procedure $\Lambda \to \infty$, we get

$$\nu_1 + \nu_{N+1} = 0.$$

Then, multiplying (1.111) by the factor $e^{\frac{2\pi i s_k}{\beta\hbar}}$ (where $k = 2, ..., N$), performing a summation over $n = 1, ..., \Lambda + 1$ and dividing the result by $\Lambda + 1$, we have

$$\frac{1}{\Lambda+1}(\nu_1 + \nu_{N+1})\frac{e^{-\frac{2\pi i(\Lambda+1)s_k}{\beta\hbar}} - 1}{e^{-\frac{2\pi i s_k}{\beta\hbar}} - 1} + \nu_k$$

$$+ \frac{1}{\Lambda+1}\sum_{\substack{2 \leqslant j \leqslant N+1 \\ j \neq k}} \nu_j \frac{e^{\frac{2\pi i(\Lambda+1)(s_j-s_k)}{\beta\hbar}} - 1}{e^{\frac{2\pi i(s_j-s_k)}{\beta\hbar}} - 1} = 0.$$

In this sum,

$$e^{\frac{2\pi i(s_j-s_k)}{\beta\hbar}} - 1 \neq 0, \quad \text{for} \quad j \neq k.$$

Thus, passing here to the limit $\Lambda \to \infty$, we get

$$\nu_k = 0, \quad k = 2, ..., N.$$

It can be seen now that $\Omega(... \nu_j ...)$ is a positive-definite quadratic form in the variables $\nu_1 + \nu_{N+1}, \nu_2, \nu_3, ..., \nu_N$. Introducing the notation

$$\nu_1' = \nu_1 + \nu_{N+1}, \quad \nu_2' = \nu_2, \quad \nu_3' = \nu_3, \quad ..., \quad \nu_N' = \nu_N, \tag{1.115}$$

we can write (1.105) in the form

$$\left\langle T\left\{\exp\left(i\sum_{1 \leqslant j \leqslant N+1} i\nu_j Q(s_j)\right)\right\}\right\rangle_\Gamma = \exp\left(-\sum_{j_1=1}^{N}\sum_{j_2=1}^{N} A_{j_1,j_2}\nu_{j_1}'\nu_{j_2}'\right), \tag{1.116}$$

where

$$\sum_{j_1=1}^{N}\sum_{j_2=1}^{N} A_{j_1,j_2}\nu_{j_1}'\nu_{j_2}' = \frac{1}{m\beta}\sum_{(n)} \frac{\left|\sum_{1 \leqslant j \leqslant N+1} e^{\frac{2\pi i n s_j}{\beta\hbar}}\nu_j'\right|^2}{(2\pi n/\beta\hbar)^2 + \omega^2} \tag{1.117}$$

is a positive-definite quadratic form in the variables $\nu_1', ..., \nu_N'$. These formulae can be used to establish the connection between averages of the T-products $\langle ... \rangle_{H^{(L)}}$ and integration in functional space. Having in mind the situation that will be studied later, we shall now consider the three-dimensional Hamiltonian

$$\Gamma = \frac{\mathbf{p}^2}{2m} + \frac{\eta^2}{2}\mathbf{r}^2 = \sum_{\alpha=1}^{3}\Gamma_\alpha, \quad \Gamma_\alpha = \frac{p_\alpha^2}{2m} + \frac{m\omega^2}{2}r_\alpha^2, \quad \omega = \frac{\eta}{m^{1/2}}. \quad (1.118)$$

Let us put

$$\mathbf{R}(s) = e^{\frac{\Gamma s}{\hbar}}\mathbf{r}e^{-\frac{\Gamma s}{\hbar}}, \quad (1.119)$$

and note that

$$R_\alpha(s) = e^{\frac{\Gamma_\alpha s}{\hbar}}r_\alpha e^{-\frac{\Gamma_\alpha s}{\hbar}} = e^{\frac{\Gamma s}{\hbar}}r_\alpha e^{-\frac{\Gamma s}{\hbar}}.$$

Because Γ_α — are operators acting upon functions of different variables, Γ_α and Γ'_α commute, and the $R_\alpha(s)$ also commute with $R_{\alpha'}(s)$ if $\alpha \neq \alpha'$. Note that

$$\langle T\{R_\alpha(s)R_{\alpha'}(s')\}\rangle_\Gamma = 0 \quad \text{for} \quad \alpha \neq \alpha'.$$

Therefore it is easily seen that the averages

$$\left\langle T\left\{\exp\left(i\sum_{1\leqslant j\leqslant N+1}\boldsymbol{\nu}_j\cdot\mathbf{R}(s_j)\right)\right\}\right\rangle_\Gamma,$$

where s_j, satisfy the condition (1.104) as before, are equal to

$$\prod_{\alpha=1}^{3}\left\langle T\left\{\exp\left(i\sum_{1\leqslant j\leqslant N+1}\nu_{j,\alpha}R_\alpha(s_j)\right)\right\}\right\rangle_{\Gamma_\alpha}.$$

Thus, taking (1.113) into account, we can write

$$\left\langle T\left\{\exp\left(i\sum_{1\leqslant j\leqslant N+1}\boldsymbol{\nu}_j\cdot\mathbf{R}(s_j)\right)\right\}\right\rangle_\Gamma$$

$$= \exp\left(-\sum_{\alpha=1}^{3}\sum_{j_1=1}^{N}\sum_{j_2=1}^{N}A_{j_1,j_2}\nu'_{j_1,\alpha}\nu'_{j_2,\alpha}\right), \quad (1.120)$$

where

$$\nu'_{1,\alpha} = \nu_{1,\alpha} + \nu_{N+1,\alpha}, \quad \nu'_{2,\alpha} = \nu_{2,\alpha}, \quad \nu'_{3,\alpha} = \nu_{3,\alpha}, \quad = \dots, \quad \nu'_{N,\alpha} = \nu_{N,\alpha}. \quad (1.121)$$

Consider the expression

$$\int\left\langle T\left\{e^{i\sum_{1\leqslant j\leqslant N+1}\boldsymbol{\nu}_j\cdot\mathbf{R}(s_j)}\right\}\right\rangle_\Gamma e^{-i\sum_{1\leqslant j\leqslant N+1}\boldsymbol{\nu}_j\cdot\mathbf{R}'_j}\,d\boldsymbol{\nu}_1\cdots d\boldsymbol{\nu}_N\,d\boldsymbol{\nu}_{N+1}, \quad (1.122)$$

where $\mathbf{R}'_j$ are real vectors and the integration with respect to each variable $\nu_{j,\alpha}$ is over the whole real axis. Let us denote

$$\frac{\nu_1 - \nu_{N+1}}{2} = \nu'_{N+1}, \qquad \nu_1 + \nu_{N+1} = \nu'_1,$$

and note that

$$\sum_{1 \leqslant j \leqslant N+1} \nu_j \cdot \mathbf{R}'_j = \nu'_1 \cdot \frac{\mathbf{R}'_1 + \mathbf{R}'_{N+1}}{2}$$

$$+ \sum_{2 \leqslant j \leqslant N} \nu_j \cdot \mathbf{R}'_j + \nu'_{N+1} \cdot (\mathbf{R}'_1 - \mathbf{R}'_{N+1}),$$

$$\int (\ldots) \, d\nu_1 \cdots d\nu_{N+1} = \int (\ldots) \, d\nu'_1 \cdots d\nu'_{N+1}.$$

From another point of view, (1.117) shows that the magnitude $\langle \ldots \rangle_\Gamma$, which is contained in the left-hand side of (1.119), does not depend on ν'_{N+1}, thus the integration over ν'_{N+1} can be done independently:

$$\int e^{-i\nu'_j(\mathbf{R}'_1 - \mathbf{R}'_{N+1})} \, d\nu'_{N+1} = (2\pi)^3 \delta(\mathbf{R}'_1 - \mathbf{R}'_{N+1}).$$

Therefore

$$\frac{1}{(2\pi)^{3(N+1)}} \int \left\langle T\left\{ e^{\left(i\sum_{1 \leqslant j \leqslant N+1} \nu_j \cdot \mathbf{R}(s_j) \right)^2} \right\} \right\rangle_\Gamma e^{-i\sum_{1 \leqslant j \leqslant N+1} \nu_j \cdot \mathbf{R}'_j} \{D_N \nu\}$$

$$= \frac{\delta(\mathbf{R}_1 - \mathbf{R}_{N+1})}{(2\pi)^{3N}} \int \prod_{\alpha=1}^{3} e^{-\sum_{j_1=1}^{N} \sum_{j_2=1}^{N} A_{j_1,j_2} \nu'_{j_1,\alpha} \nu'_{j_2,\alpha} - i\sum_{j=1}^{N} \nu'_{j,\alpha} R'_{j,\alpha}} \{D_N \nu'\}$$

$$= \frac{\delta(\mathbf{R}_1 - \mathbf{R}_{N+1})}{(2\pi)^{3N}} \prod_{\alpha=1}^{3} \int e^{-\sum_{j_1=1}^{N} \sum_{j_2=1}^{N} A_{j_1,j_2} x_{j_1} x_{j_2} - i\sum_{j=1}^{N} x_j R'_{j,\alpha}} \{D_N x\}.$$

where $\{D_{N+1} \nu\} = d\nu_1 \cdots d\nu_N \, d\nu_{N+1}$, $\{D_N \nu'\} = d\nu'_{1,\alpha} d\nu'_{2,\alpha} \cdots d\nu'_{N,\alpha}$, a $\{D_N x\} = dx_1 \, dx_2 \cdots dx_N$. Calculation of the usual Gaussian integral results in

$$\int \exp\left(-\sum_{j_1=1}^{N} \sum_{j_2=1}^{N} A_{j_1,j_2} x_{j_1} x_{j_2} - i\sum_{j=1}^{N} x_j R'_{j,\alpha} \right) dx_1 \, dx_2 \cdots dx_N$$

$$= \frac{\pi^{N/2}}{(\text{Det} A)^{1/2}} \exp\left(-\frac{1}{4} \sum_{j_1=1}^{N} \sum_{j_2=1}^{N} (A^{-1})_{j_1,j_2} R'_{j_1,\alpha} R'_{j_2,\alpha} \right).$$

Because the matrix A is positive-definite,

$$\mathrm{Det}\, A > 0,$$

the inverse matrix A^{-1} is also positive-definite. We now have

$$\frac{1}{(2\pi)^{3(N+1)}} \int \left\langle T\left\{ e^{i\sum_{1\leqslant j\leqslant N+1} \boldsymbol{\nu}_j \cdot \mathbf{R}(s_j)} \right\} \right\rangle_\Gamma e^{-i\sum_{1\leqslant j\leqslant N+1} \boldsymbol{\nu}_j \cdot \mathbf{R}'_j} \{D_{N+1}\boldsymbol{\nu}\}$$
$$= \rho(\mathbf{R}'_1, \mathbf{R}'_2, ..., \mathbf{R}'_{N+1}), \quad (1.123)$$

where $\{D_{N+1}\boldsymbol{\nu}\} = d\boldsymbol{\nu}_1 \cdots d\boldsymbol{\nu}_N\, d\boldsymbol{\nu}_{N+1}$ and

$$\rho(\mathbf{R}'_1, \mathbf{R}'_2, ..., \mathbf{R}'_{N+1}) = \frac{1}{2^{3N}\pi^{3N/2}} \frac{\delta(\mathbf{R}'_1 - \mathbf{R}'_{N+1})}{(\mathrm{Det}\, A)^{3/2}}$$

$$\times \exp\left(-\frac{1}{4} \sum_{j_1=1}^{N} \sum_{j_2=1}^{N} (A^{-1})_{j_1,j_2} R'_{j_1,\alpha} R'_{j_2,\alpha} \right). \quad (1.124)$$

It is obvious that

$$\rho(\mathbf{R}_1, \mathbf{R}_2, ..., \mathbf{R}_{N+1}) \geqslant 0,$$
$$\int \rho(\mathbf{R}_1, \mathbf{R}_2, ..., \mathbf{R}_{N+1})\, d\mathbf{R}_1\, d\mathbf{R}_2 \cdots d\mathbf{R}_{N+1} = 1. \qquad (1.125)$$

These equations can be applied to disentangle expressions of the kind

$$\langle T\{f(\mathbf{R}(s_1), ..., \mathbf{R}(S_{N+1})\}\rangle_\Gamma.$$

So, consider a function

$$f(\mathbf{R}_1, ..., \mathbf{R}_{N+1}),$$

that depends on $N + 1$ real vectors and whose Fourier representation is

$$f(...\mathbf{R}_j...)$$
$$= \frac{1}{(2\pi)^{3(N+1)}} \int e^{i\left(\sum_{(j)} \boldsymbol{\nu}_j \mathbf{R}_j - \sum_{(j)} \boldsymbol{\nu}_j \mathbf{R}'_j\right)} f(...\mathbf{R}'_j...) \{D_{N+1}\mathbf{R}'\}\{D_{N+1}\boldsymbol{\nu}\},$$

where $j = 1, ..., N + 1$ and $\{D_{N+1}\mathbf{R}'\} = d\mathbf{R}'_1 d\mathbf{R}'_2 \cdots d\mathbf{R}'_{N+1}$. Because the operators $\mathbf{R}(s_j)$ commute under the sign of the T-product, we can write

$$T\{f(\mathbf{R}(s_1), ..., \mathbf{R}(s_{N+1}))\} = \frac{1}{(2\pi)^{3(N+1)}}$$

$$\times \int T\left\{ e^{i\sum_{(j)} \mathbf{R}(s_j)\cdot\boldsymbol{\nu}_j} \right\} e^{-i\sum_{(j)} \boldsymbol{\nu}_j \cdot \mathbf{R}'_j} f(...\mathbf{R}'_j...) \{D_{N+1}\mathbf{R}'\}\{D_{N+1}\boldsymbol{\nu}\},$$

where $\{D_{N+1}\mathbf{R'}\} = d\mathbf{R'_1}\,d\mathbf{R'_2}\cdots d\mathbf{R'_{N+1}}$. Therefore, thanks to (1.120),

$$\langle T\{f(\mathbf{R}(s_1),...,\mathbf{R}(S_{N+1})\}\rangle_\Gamma$$
$$= \int f(...\mathbf{R'_j}...)\rho(...\mathbf{R'_j}...)\,d\mathbf{R'_1}\,d\mathbf{R'_2}\cdots d\mathbf{R'_{N+1}}, \quad (1.126)$$

and, taking (1.122) into account, we see that

$$M \geqslant \langle T\{f(...\mathbf{R}(s_j)...)\}\rangle_\Gamma \geqslant 0,$$

if

$$M \geqslant f(...\mathbf{R}(s_j)...) \geqslant 0. \quad (1.127)$$

Now let us consider functionals $F(\mathbf{R})$ that depend on real functions $\mathbf{R}(s)$, defined on the interval

$$0 \leqslant s \leqslant \beta\hbar,$$

and use the following notation for the functional integral:

$$\int F(\mathbf{R})\,d\mu.$$

For a subset of "special functionals" of the type

$$F(\mathbf{R}) = \Phi(...\mathbf{R}(s_j)...),$$

that depend on a finite number N of vectors $\mathbf{R}(s_j)$, we define this integral by the following procedure:

$$\int F(\mathbf{R})\,d\mu = \int \Phi(...\mathbf{R}(s_j)...)\rho(...\mathbf{R}_j...) \prod_{(j)} d\mathbf{R}_j.$$

Then, thanks to (1.123) and (1.124),

$$\langle T\{f(\mathbf{R})\}\rangle_\Gamma = \int F(\mathbf{R})\,d\mu,$$
$$\langle T\{f(\mathbf{R})\}\rangle_\Gamma \geqslant 0 \quad \text{if } F(\mathbf{R}) \geqslant 0 \quad (1.128)$$

for arbitrary real vectors $\mathbf{R}(s)$. These relations can be generalized for a broader set of functionals $F(\mathbf{R})$ if one approximates these functionals by corresponding sequences of the above-mentioned "special functionals" with subsequent passage to the limit $N \to \infty$. For example, we can consider a sequence of functions

$$\mathbf{R}_N(s) = \mathbf{R}(s_j), \quad s_j \leqslant s \leqslant s_{j+1}, \quad j = 1,...,N \quad (1.129)$$

$$s_1 = 0, \quad s_{N+1} = \beta\hbar,$$

$$|s_{j+1} - s_j| \leqslant \Delta s \to 0, \quad N \to \infty$$

and approximate the functional $F(\mathbf{R})$ in question by the form $F(\mathbf{R}_N)$, belonging obviously to the class of "special functionals". We should like to stress here that the technique of functional integration was developed first by R. Feynman. His method (known as the "path integration method") was also applied by him to problems of quantum statistical mechanics [5, 39].

In our approach, we prefer to deal directly with the averaged T-products

$$\langle T\{F(\mathbf{R})\}\rangle_\Gamma.$$

Moreover, the only property we require is the nonnegativity of all these averages in the case in which the functionals $F(\mathbf{R})$ are positive:

$$\langle T\{F(\mathbf{R})\}\rangle_\Gamma \geqslant 0, \tag{1.130}$$

if $F(\mathbf{R}) \geqslant 0$ for arbitrary real vectors $\mathbf{R}(s)$. Let us choose instead of the Hamiltonian Γ, given by expression (1.83), the Hamiltonian

$$H(\Sigma) = \frac{1}{2}\sum_{(f)}(p_f p_f{}^\dagger + \omega^2 q_f q_f^\dagger), \tag{1.131}$$

$$p_f^\dagger = p_{-f}, \quad q_f^\dagger = q_{-f}.$$

As has already been stressed, one can introduce Bose amplitudes by

$$q_f = \left(\frac{\hbar}{2\omega(f)}\right)^{1/2}(b_f + b_{-f}^\dagger), \qquad p_f = i\left(\frac{2\omega(f)}{\hbar}\right)^{1/2}(b_{-f}^\dagger - b_f), \tag{1.132}$$

and can transform $H(\Sigma)$ into the form

$$H(\Sigma) = \sum_{(f)}\hbar\omega(f)b_f^\dagger b_f + \frac{1}{2}\sum_{(f)}\hbar\omega(f). \tag{1.133}$$

Let us consider

$$\left\langle T\left\{e^{\int_0^{\beta\hbar} ds \sum_{(f)} \Lambda_f(s)Q_f(s)}\right\}\right\rangle_{H(\Sigma)} = \frac{\mathrm{Tr}\, e^{-\beta H(\Sigma)}T\{e^{\int_0^{\beta\hbar} ds \sum_{(f)} \Lambda_f(s)Q_f(s)}\}}{\mathrm{Tr}\, e^{-\beta H(\Sigma)}}, \tag{1.134}$$

where

$$Q_f(s) = \exp\left(\frac{sH(\Sigma)}{\hbar}\right)q_f\exp\left(-\frac{sH(\Sigma)}{\hbar}\right). \tag{1.135}$$

Here we are going to analyze the situation where all operators $\Lambda(f)$ commute with each other, as well as with all operators $Q_f(s)$, $H(\Sigma)$, so that when calculating (1.131) we can treat $\Lambda_f(s)$ as the usual C-functions. Hence, using (1.82), we get

$$\left\langle T\left\{e^{\int\limits_0^{\beta\hbar} ds \sum\limits_{(f)} \Lambda_f(s)Q_f(s)}\right\}\right\rangle_{H(\Sigma)}$$

$$\exp\left(\frac{1}{2}\int\limits_0^{\hbar\beta} ds \int\limits_0^{\hbar\beta} d\sigma \left\langle T\left\{\sum_{(f)}\Lambda_f(s)Q_f(s)\sum_{(f')}\Lambda_{f'}Q_{f'}(\sigma)\right\}\right\rangle_{H(\Sigma)}\right).$$

On the other hand, it follows from (1.89), (1.129) and (1.130) that

$$\left\langle T\left\{\sum_{(f)}\Lambda_f(s)Q_f(s)\sum_{f'}\Lambda_{f'}Q_{f'}(\sigma)\right\}\right\rangle_{H(\Sigma)}$$

$$= \sum_{(f)}\Lambda_f(s)\Lambda_{-f}(\sigma)\langle T\{Q_f(s)Q_{-f}(\sigma)\}\rangle_{H(\Sigma)}$$

$$= \sum_{(f)}\Lambda_f(s)\Lambda_{-f}(\sigma)\frac{\hbar}{2\omega_f}\left(1 - e^{-\beta\omega_f\hbar}\right)^{-1}\left(e^{-\omega_f|s-\sigma|} + e^{-\omega_f\beta\hbar+\omega_f|s-\sigma|}\right).$$

Thus

$$\left\langle T\left\{e^{\int\limits_0^{\beta\hbar} ds \sum\limits_{(f)} \Lambda_f(s)Q_f(s)}\right\}\right\rangle_{H(\Sigma)}$$

$$= \exp\left(\int\limits_0^{\hbar\beta} ds \int\limits_0^{\hbar\beta} d\sigma \sum_{(f)} \frac{\hbar\Lambda_f(s)\Lambda_{-f}(\sigma)}{4\omega_f}\frac{(e^{-\omega_f|s-\sigma|} + e^{-\omega_f\beta\hbar+\omega_f|s-\sigma|})}{1 - e^{-\beta\omega_f\hbar}}\right).$$

$$(1.136)$$

Let us assume that $...\Lambda_f...$ are operators commuting with any of the operators $...Q...$ and $H(\Sigma)$, but not commuting with each other. Of course, equation (1.133) does not hold at all in this case. However, consider the situation where the left-hand side of (1.133) is subjected to another T-ordering operation that does not affect the operators Q_f, $H(\Sigma)$, but puts in order only operators containing $...\Lambda_f...$. In short, we consider just the expression

$$T'\left\{\langle T\{e^{\int\limits_0^{\beta\hbar} ds[A(s)+\sum\limits_{(f)}\Lambda_f(s)Q_f(s)]}\}\rangle_{H(\Sigma)}\right\}, \qquad (1.137)$$

where $A,...\Lambda_f...$ are operators dependent only on those variables of the wave function that are not influenced by the operators $H(\Sigma),...Q_f...$ and vice versa. Here the symbol T' implies only the procedure of ordering for operators

$$A, ... \Lambda_f ...$$

Then (1.134) can be rewritten directly with the help of (1.133), as if $A, \ldots \Lambda_f \ldots$ were the usual C-functions. Consequently,

$$
T'\left\{\left\langle T\left\{e^{\int_0^{\beta\hbar} ds[A(s)+\sum_{(f)}\Lambda_f(s)Q_f(s)]}\right\}\right\rangle_{H(\Sigma)}\right\}
$$

$$
= T'\left\{\exp\left(\frac{1}{2}\int_0^{\hbar\beta}ds\int_0^{\hbar\beta}d\sigma\sum_{(f)}\Lambda_f(s)\Lambda_{-f}(\sigma)\frac{\hbar}{2\omega_f}\left(1-e^{-\beta\omega_f\hbar}\right)^{-1}\right.\right.
$$

$$
\left.\left.\times\left(e^{-\omega_f|s-\sigma|}+e^{-\omega_f\beta\hbar+\omega_f|s-\sigma|}\right)+\int_0^{\hbar\beta}ds\,A(s)\right)\right\}. \quad (1.138)
$$

1.7. Auxiliary Operator Identities

Let us consider the operator equation

$$
\hbar\frac{dU(s)}{ds} = -\{H_0 + H_1(s)\}U(s), \quad (1.139)
$$

where $U(0) = \widehat{1}$ — is the unit operator, and $H_1(s)$ is an operator that can depend explicitly on s. It is easily seen that

$$
U(s) = T\left\{e^{-\frac{1}{\hbar}\int_0^s\{H_0+H_1(\sigma)\}\,d\sigma}\right\}.
$$

Then substitute into (1.136)

$$
U(s) = e^{-\frac{H_0 s}{\hbar}}C(s).
$$

This ansatz leads to the following equation for $C(s)$

$$
\hbar\frac{dC(s)}{ds} = -e^{\frac{H_0 s}{\hbar}}H_1(s)e^{-\frac{H_0 s}{\hbar}}C(s), \quad C(0) = 1,
$$

which equation can be solved formally as

$$
C(s) = T\left\{e^{-\frac{1}{\hbar}\int_0^s d\sigma\,e^{\frac{H_0\sigma}{\hbar}}H_1(\sigma)e^{-\frac{H_0\sigma}{\hbar}}}\right\}.
$$

Therefore

$$
U(\beta\hbar) = T\left\{e^{-\frac{1}{\hbar}\int_0^{\beta\hbar}ds\,\{H_0(s)+H_1(s)\}}\right\}
$$

$$
= e^{-H_0\beta}T\left\{e^{-\frac{1}{\hbar}\int_0^{\beta\hbar}ds\,e^{\frac{H_0 s}{\hbar}}H_1(s)e^{-\frac{H_0 s}{\hbar}}}\right\}. \quad (1.140)
$$

In the particular case, where $H_1(s) = H_1$ and does not depend on s explicitly, (1.136) gives

$$U(s) = e^{-(H_0+H_1)\frac{s}{\hbar}},$$

and it follows from (1.137) that

$$e^{-\beta(H_0+H_1)} = e^{-\beta H_0} T \left\{ e^{-\frac{1}{\hbar} \int\limits_{0}^{\beta\hbar} ds\, e^{\frac{H_0 s}{\hbar}} H_1 e^{-\frac{H_0 s}{\hbar}}} \right\}. \qquad (1.141)$$

Chapter 2

EQUILIBRIUM THERMODYNAMIC STATE
OF POLARON SYSTEM

The main objective of this chapter is the derivation of Bogolubov's inequality for the polaron reduced free energy by means of the algebraic T-product method. This inequality is a source of various upper bounds for the polaron ground-state energy. Feynman's well-known inequality is reproduced as a particular case of Bogolubov's inequality. A weak-interaction systematic finite-temperature perturbation scheme, based on the same T-product formalism, as well as the adiabatic perturbation approach, valid for the strong-coupling case, are also outlined.

2.1. Free Energy and Ground State Energy Calculation

Let us consider the standard polaron Hamiltonian

$$H_P = H(S) + H(\Sigma) + H_{\text{int}}(S, \Sigma), \qquad (2.1)$$

where

$$H(S) = \frac{\mathbf{p}^2}{2m},$$

$$H(\Sigma) = \frac{1}{2} \sum_{(f)} \{p_f p_f^\dagger + \omega^2(f) q_f q_f^\dagger\}, \quad q_{-f} = q_f^\dagger, \quad p_{-f} = p_f^\dagger, \qquad (2.2)$$

$$H_{\text{int}}(S, \Sigma) = \frac{1}{V^{1/2}} \sum_{(f)} L(f) q_f e^{i\mathbf{f} \cdot \mathbf{r}}, \quad L(f) = L(-f) = L^\dagger(f).$$

Here, as in Chapter 1,

$$f = \left(\frac{2\pi n_1}{L}, \frac{2\pi n_2}{L}, \frac{2\pi n_3}{L} \right),$$

where (n_1, n_2, n_3) are arbitrary integers and $L^3 = V$.

We can see that $H(S)$ is the Hamiltonian of a free particle of mass m, (the electron in polaron theory); $H(\Sigma)$ is the Hamiltonian of the phonon field, and the Hamiltonian $H_{\text{int}}(S, \Sigma)$ describes the interaction between the two systems Σ and S. In the case of the standard Fröhlich model

$$L(f) = \frac{\mathrm{g}}{|f|}, \quad \mathrm{g} = \text{const}, \qquad (2.3)$$

$$\omega(f) = \omega = \text{const}.$$

If we want the Hamiltonian $H_{\text{int}}(S, \Sigma)$ to contain only a finite number of terms (until the formal passage to the limit $V \to \infty$), we must put

$$
L(f) = \begin{cases} \dfrac{g}{|f|} & \text{for} \quad |f| \leqslant f_{\max} \\ 0 & \text{for} \quad |f| \geqslant f_{\max} \end{cases} \tag{2.4}
$$

Staying within the framework of the standard Fröhlich model, we shall assume that

$$f_{\max} \to \infty \quad \text{when} \quad V \to \infty.$$

However, we can consider the case when $f_{\max}$ is held fixed forever during passage to the thermodynamic limit. The physical justification for this assumption originates from the fact that the Fröhlich model does not take proper account of the lattice structure. Thus a contribution to the interaction Hamiltonian $H_{\text{int}}(S, \Sigma)$ from any vector f such that $|f| > 2\pi/a$, where a is the lattice constant, is not represented correctly in the expression (2.2) and should be omitted.

The free energy for the Hamiltonian H_P is given by the usual expression

$$-\vartheta \ln \text{Tr}_{S,\Sigma}\, e^{\beta H_P}.$$

This quantity is divergent in general. To make it finite, we must subtract the free energies corresponding to the free particle S and to the free phonon field Σ. As a result, we arrive at the so-called interaction free energy corresponding to the Hamiltonian $H_{\text{int}}(S, \Sigma)$:

$$
-\vartheta \ln \text{Tr}_{S,\Sigma}\, e^{\beta H_P} - \left(-\vartheta \ln \text{Tr}_S\, e^{\beta H(S)} - \vartheta \ln \text{Tr}_\Sigma\, e^{\beta H(\Sigma)} \right)
$$
$$
= -\vartheta \ln \frac{\text{Tr}_{S,\Sigma}\, e^{-\beta H_P}}{\text{Tr}_S\, e^{-\beta H(S)}\, \text{Tr}_\Sigma\, e^{-\beta H(\Sigma)}}. \tag{2.5}
$$

Later we shall keep this expression in mind when considering the polaron free energy.

It is usually assumed that the particle in question (i.e. the electron in polaron theory) is confined within a limited region of space. This means that the radius vector $\mathbf{r}$ of the particle S lays inside a finite volume V. However, from a technical point of view, it is far more convenient to assume that the radius vector $\mathbf{r}$ can take any value. To compensate for the unwanted consequences of this assumption (namely the divergence of the free energy of the free particle), we add an auxiliary term $\eta^2 \mathbf{r}^2/2$ to the Hamiltonian $H(S)$, which now reads

$$
H(S) = \Gamma = \frac{\mathbf{p}^2}{2m} + \frac{\eta^2 \mathbf{r}^2}{2}. \tag{2.6}
$$

This auxiliary term ensures soft confinement of the particle within some effective volume. Smaller values of η lead to softer confinement, and thus to larger effective volume, and vice versa. Of course, we take the passage to the thermodynamic limit

$$\eta \to \infty, \quad V \to \infty. \tag{2.7}$$

in the final results. In so doing, we see that the interaction free energy can be represented finally by the following expression:

$$F_{\text{int}}^{(p)} = \lim_{\eta \to 0} \lim_{V \to \infty} F_{\text{int}}^{(p)}(V, \eta),$$

where

$$F_{\text{int}}^{(p)}(V, \eta) = -\vartheta \ln \frac{\text{Tr}_{S,\Sigma} \, e^{-\beta H_P}}{\text{Tr}_S \, e^{-\beta H(S)} \, \text{Tr}_\Sigma \, e^{-\beta H(\Sigma)}}. \tag{2.8}$$

Let us use (1.138), in which we choose

$$H_0 = H(S) + H(\Sigma) = \Gamma + H(\Sigma), \tag{2.9}$$

$$H_1 = H_{\text{int}}(S, \Sigma).$$

We denote

$$\mathbf{R}(s) = e^{\frac{H_0 s}{\hbar}} \mathbf{r} e^{-\frac{H_0 s}{\hbar}}, \qquad Q_f(s) = e^{\frac{H_0 s}{\hbar}} q_f e^{-\frac{H_0 s}{\hbar}}.$$

Then (2.2) results in

$$e^{\frac{H_0 s}{\hbar}} H_{\text{int}}(S, \Sigma) e^{-\frac{H_0 s}{\hbar}} = \frac{1}{V^{1/2}} \sum_{(f)} L(f) Q_f(s) e^{i f \cdot \mathbf{R}(s)},$$

and it follows from (1.138) that

$$e^{-\beta H_P} = e^{-\beta H(S)} e^{-\beta H(\Sigma)} T \left\{ \exp\left(-\frac{1}{\hbar V^{1/2}} \int\limits_0^{\hbar\beta} ds \sum_{(f)} L(f) Q_f(s) e^{i f \cdot \mathbf{R}(s)} \right) \right\}. \tag{2.10}$$

Note that $H(S) = \Gamma$ and $H(\Sigma)$ act in completely different subspaces, corresponding to the systems S and Σ. Therefore

$$\mathbf{R}(s) = e^{\frac{\Gamma}{\hbar} s} \mathbf{r} e^{-\frac{\Gamma}{\hbar} s}, \qquad Q_f(s) = e^{\frac{H(\Sigma)}{\hbar} s} q_f e^{-\frac{H(\Sigma)}{\hbar} s}. \tag{2.11}$$

Because $\mathbf{R}(s)$ acts only upon the variables of the system S, and $Q_f(s)$ only upon the variables of the system Σ, the ordering T-operation in (2.10) can be carried out in two steps: first of all we put into the proper order all the operators $Q_f(s)$, and after this we order all the operators $\mathbf{R}(s)$. Thus we find from (2.10) that

$$e^{-\beta H_P} = e^{-\beta \Gamma} e^{-\beta H(\Sigma)} T_{\mathbf{R}} \left\{ T_{\mathbf{Q}} \left\{ \exp\left(-\frac{1}{\hbar V^{1/2}} \int\limits_0^{\hbar\beta} ds \sum_{(f)} L(f) Q_f(s) e^{i f \cdot \mathbf{R}(s)} \right) \right\} \right\}. \tag{2.12}$$

Note further that the operator $e^{\beta H(\Sigma)}$ can be placed after the symbol denoting the operation $T_{\mathbf{R}}$, because this operator does not act upon the variables $\mathbf{R}(s)$. Taking this possibility into account, we find from (2.8)

$$F_{\text{int}}^{(P)}(V,\eta) = -\vartheta \ln \frac{\text{Tr}_S\, e^{-\beta\Gamma} e^{-\beta H(\Sigma)} T_{\mathbf{R}}\{F(\mathbf{R})\}}{\text{Tr}_{(S)}\, e^{-\beta\Gamma}}, \qquad (2.13)$$

where

$$F(\mathbf{R}) = \left\langle T_{\mathbf{Q}}\left\{\exp\left(-\frac{1}{\hbar V^{1/2}}\int_0^{\hbar\beta} ds \sum_{(f)} L(f) Q_f(s) e^{i\mathbf{f}\cdot\mathbf{R}(s)}\right)\right\}\right\rangle_{H(\Sigma)}.$$

In order to transform the right-hand side of this expression, we can use (1.134), in which we make the substitution

$$T' \to T_{\mathbf{R}}, \quad T \to T_{\mathbf{Q}}$$

and put

$$A(s) = 0, \quad \Lambda_f(S) = \frac{1}{\hbar V^{1/2}} L(f) e^{i\mathbf{f}\cdot\mathbf{R}(s)}, \quad \omega_f = \omega. \qquad (2.14)$$

If we observe that

$$L(f)L(-f) = |L(f)|^2,$$

we obtain

$$T_{\mathbf{R}}\{e^{\Phi}\} = T_{\mathbf{R}}\left\{\left\langle T_{\mathbf{Q}}\left\{\exp\left(-\frac{1}{\hbar V^{1/2}}\int_0^{\hbar\beta} ds \sum_{(f)} L(f) Q_f(s) e^{i\mathbf{f}\cdot\mathbf{R}(s)}\right)\right\}\right\rangle_{H(\Sigma)}\right\},$$
$$(2.15)$$

where

$$\Phi = \frac{1}{2\hbar^2 V} \sum_{(f)} |L(f)|^2 \frac{\hbar}{2\omega} (1 - e^{-\beta\hbar\omega})^{-1}$$

$$\times \int_0^{\hbar\beta} ds \int_0^{\hbar\beta} d\sigma (e^{-\omega|s-\sigma|} + e^{-\omega\beta\hbar + \omega|s-\sigma|}) e^{i\mathbf{f}\cdot\{\mathbf{R}(s)-\mathbf{R}(\sigma)\}}. \qquad (2.16)$$

Since e^{Φ} on the right-hand side of (2.15) is a functional of the variables $\mathbf{R}(s)$ only, not containing the variables $Q_f(s)$, there is no need to introduce special notation $T_{\mathbf{R}}$. Therefore we shall denote the right-hand side of (2.15) simply as

$$T\{e^{\Phi}\}.$$

Thus we obtain from (2.13) that

$$F_{\text{int}}^{(P)}(V,\eta) = -\vartheta \ln \langle T\{e^{\Phi}\}\rangle_{\Gamma}. \qquad (2.17)$$

Let us return now to the quadratic Hamiltonian of Chapter 1 and rewrite it in the form

$$H^{(L)} = H(S) + H^{(L)}(\Sigma) + H^L_{\text{int}}(S, \Sigma), \qquad (2.18)$$

where

$$H(S) = \Gamma$$

is given by the expression (2.6) as before, and

$$H^{(L)}(\Sigma) = \frac{1}{2} \sum_{(f)} \{p_f p_f^\dagger + \nu^2(f) q_f q_f^\dagger\},$$

$$H^{(L)}_{\text{int}}(S, \Sigma) = \frac{K_0^2 \mathbf{r}^2}{2} + \frac{i}{V^{1/2}} \sum_{(f)} S(f) q_f \, \mathbf{f} \cdot \mathbf{r} \qquad (2.19)$$

$$= \frac{1}{6V} \sum_{(f)} \frac{S^2(f) f^2}{\nu^2(f)} \mathbf{r}^2 + \frac{i}{V^{1/2}} \sum_{(f)} S(f) q_f \, \mathbf{f} \cdot \mathbf{r}.$$

We can repeat for the Hamiltonian $H^{(L)}$ all of our previous arguments, developed for the Hamiltonian $H^{(P)}$, if we put in (1.135)

$$A(s) = -\frac{K_0^2 \mathbf{R}^2(s)}{2\hbar}, \qquad \Lambda_f(S) = -\frac{i}{\hbar V^{1/2}} \sum_{(f)} S(f) \, \mathbf{f} \cdot \mathbf{R}(s).$$

As a result, we obtain, instead of (2.17), the new expression

$$F^{(L)}_{\text{int}}(V, \eta) = -\vartheta \ln \frac{\text{Tr}_{S,\Sigma} \, e^{\beta H^{(L)}}}{\text{Tr}_S \, e^{\beta \Gamma} \, \text{Tr}_\Sigma \, e^{\beta H^{(L)}(\Sigma)}} = -\vartheta \ln \langle T\{e^{\Phi_0}\}\rangle_\Gamma, \qquad (2.20)$$

where

$$\Phi_0 = -\frac{K_0^2}{2\hbar} \int\limits_0^{\hbar\beta} ds \, \mathbf{R}^2(s) + \frac{1}{2\hbar^2 V} \int\limits_0^{\hbar\beta} ds \int\limits_0^{\hbar\beta} d\sigma \sum_{(f)} S^2(f) \frac{\hbar}{2\nu(f)}$$

$$\times \frac{K_{\nu(f)}(s, \sigma) \mathbf{f} \cdot \mathbf{R}(\sigma)}{1 - e^{-\beta\hbar\nu(f)}} = -\frac{K_0^2}{2\hbar} \int\limits_0^{\hbar\beta} ds \, \mathbf{R}^2(s)$$

$$+ \frac{1}{2\hbar^2 V} \int\limits_0^{\hbar\beta} ds \int\limits_0^{\hbar\beta} d\sigma \sum_{(f)} S^2(f) \frac{\hbar f^2}{6\nu(f)} \frac{K_{\nu(f)}(s, \sigma) \mathbf{R}(s) \cdot \mathbf{R}(\sigma)}{1 - e^{-\beta\hbar\nu(f)}}, \qquad (2.21)$$

and

$$K_{\nu(f)}(s, \sigma) = e^{-\nu(f)|s-\sigma|} + e^{-\nu(f)\beta\hbar + \nu(f)|s-\sigma|}. \qquad (2.22)$$

It follows from (1.91) and (1.102) that

$$K_{\nu(f)}(s,\sigma) = 2\nu(f)\frac{1 - e^{-\beta\hbar\nu(f)}}{\beta\hbar} \sum_{(n)} \frac{e^{2\pi i n \frac{s-\sigma}{\beta\hbar}}}{(2\pi n/\beta\hbar)^2 + \nu^2(f)}.$$

Hence

$$\int_0^{\hbar\beta} d\sigma \frac{\hbar}{\nu(f)} \left(1 - e^{-\beta\hbar\nu(f)}\right)^{-1} K_{\nu(f)}(s,\sigma)$$

$$= \int_0^{\hbar\beta} ds \frac{\hbar}{\nu(f)} \left(1 - e^{-\beta\hbar\nu(f)}\right)^{-1} K_{\nu(f)}(s,\sigma) = \frac{2}{\beta} \frac{\beta\hbar}{\nu^2(f)} = \frac{2\hbar}{\nu^2(f)}.$$

Therefore

$$\frac{1}{4\hbar^2} \int_0^{\hbar\beta} ds \int_0^{\hbar\beta} d\sigma \frac{\hbar}{6\nu(f)} \left(1 - e^{-\beta\hbar\nu(f)}\right)^{-1} K_{\nu(f)}(s,\sigma)[\mathbf{R}^2(s) + \mathbf{R}^2(\sigma)]$$

$$= \frac{1}{4\hbar} \int_0^{\hbar\beta} ds \frac{1}{3\nu^2(f)} \mathbf{R}^2(s) + \frac{1}{4\hbar} \int_0^{\hbar\beta} d\sigma \frac{1}{3\nu^2(f)} \mathbf{R}^2(\sigma) = \frac{1}{2\hbar} \int_0^{\hbar\beta} ds \frac{1}{3\nu^2(f)} \mathbf{R}^2(s).$$

Since

$$K_0^2 = \frac{1}{V} \sum_{(f)} \frac{S^2(f)}{\nu^2(f)} \frac{|f|^2}{3},$$

we can see that

$$\frac{K_0^2}{2\hbar} \int_0^{\hbar\beta} ds \mathbf{R}^2(s) = \frac{1}{4\hbar^2 V} \int_0^{\hbar\beta} ds \int_0^{\hbar\beta} d\sigma \sum_{(f)} \frac{S^2(f)f^2}{6\nu(f)} \hbar \frac{K_{\nu(f)}(s,\sigma)[\mathbf{R}^2(s) + \mathbf{R}^2(\sigma)]}{1 - e^{-\beta\hbar\nu(f)}}.$$

If we introduce the notation

$$L(|s-\sigma|) = \frac{1}{4\hbar^2 V} \sum_{(f)} \frac{S^2(f)f^2}{6\nu(f)} \hbar \frac{[e^{-\nu(f)|s-\sigma|} + e^{-\nu(f)\beta\hbar + \nu(f)|s-\sigma|}]}{(1 - e^{-\beta\hbar\nu(f)})}. \quad (2.23)$$

then it follows from (2.21) that

$$\Phi_0 = -\int_0^{\hbar\beta} ds \int_0^{\hbar\beta} d\sigma \, L(|s-\sigma|)[\mathbf{R}(s) - \mathbf{R}(\sigma)]^2. \quad (2.24)$$

As the next step, we shall try to consider this functional as a possible approximation to the true initial functional Φ. Our aim is to derive an approximate expression for the polaron interaction free energy (2.17).

Generally speaking, if y is a quantity of the first order of smallness then, up to terms of the second order of smallness (in fact, we may neglect such terms)

$$\ln(x+y) - \ln(x) = \int\limits_{x}^{x+y} \frac{d\xi}{\xi} \simeq \frac{y}{x}.$$

Thus, regarding the difference $\Phi - \Phi_0$ as an expression of the first order of smallness from a formal point of view, we can write in the "first approximation"

$$\ln\langle T\{e^{\Phi}\}\rangle_\Gamma = \ln\langle T\{e^{\Phi_0}\}\rangle_\Gamma + \frac{\langle T\{e^{\Phi}(\Phi - \Phi_0)\}\rangle_\Gamma}{\langle T\{e^{\Phi_0}\}\rangle_\Gamma}.$$

It is interesting to note that the corresponding approximation

$$\text{app}F_{\text{int}}^{(P)}(V, \eta) = -\vartheta \ln\langle T\{e^{\Phi_0}\}\rangle_\Gamma - \vartheta\frac{\langle T\{e^{\Phi}(\Phi - \Phi_0)\}\rangle_\Gamma}{\langle T\{e^{\Phi_0}\}\rangle_\Gamma} \qquad (2.25)$$

provides us with an upper bound on the interaction free energy calculated for the initial polaron Hamiltonian (here "app" assumes approximation for precise linear model):

$$\text{app}\, F_{\text{int}}^{(P)}(V, \eta) \geqslant F_{\text{int}}^{(P)}(V, \eta). \qquad (2.26)$$

To prove this statement, consider the function

$$f(\xi) = \ln\langle T\{e^{\Phi_0 + \xi(\Phi - \Phi_0)}\}\rangle_\Gamma. \qquad (2.27)$$

We have

$$f'(\xi) = \frac{\langle T\{e^{\Phi_0 + \xi(\Phi - \Phi_0)}(\Phi - \Phi_0)\}\rangle_\Gamma}{\langle T\{e^{\Phi_0 + \xi(\Phi - \Phi_0)}\}\rangle_\Gamma}, \qquad (2.28)$$

and hence

$$\frac{\langle T\{e^{\Phi_0 + \xi(\Phi - \Phi_0)}(\Phi - \Phi_0 - f'(\xi))\}\rangle_\Gamma}{\langle T\{e^{\Phi_0 + \xi(\Phi - \Phi_0)}\}\rangle_\Gamma} = 0. \qquad (2.29)$$

Further, by means of (2.28) we find that

$$f''(\xi) = \frac{\langle T\{e^{\Phi_0 + \xi(\Phi - \Phi_0)}(\Phi - \Phi_0)^2\}\rangle_\Gamma}{\langle T\{e^{\Phi_0 + \xi(\Phi - \Phi_0)}\}\rangle_\Gamma}$$

$$- \frac{\langle T\{e^{\Phi_0 + \xi(\Phi - \Phi_0)}(\Phi - \Phi_0)\}\rangle_\Gamma \langle T\{e^{\Phi_0 + \xi(\Phi - \Phi_0)}(\Phi - \Phi_0)\}\rangle_\Gamma}{\langle T\{e^{\Phi_0 + \xi(\Phi - \Phi_0)}\}\rangle_\Gamma^2}.$$

It follows from this equation that

$$f''(\xi) = \frac{\langle T\{e^{\Phi_0 + \xi(\Phi - \Phi_0)}(\Phi - \Phi_0)^2\}\rangle_\Gamma}{\langle T\{e^{\Phi_0 + \xi(\Phi - \Phi_0)}\}\rangle_\Gamma} = -(f'(\xi))^2$$

$$= \frac{\langle T\{e^{\Phi_0 + \xi(\Phi - \Phi_0)}[(\Phi - \Phi_0)^2 - (f'(\xi))^2)]\}\rangle_\Gamma}{\langle T\{e^{\Phi_0 + \xi(\Phi - \Phi_0)}\}\rangle_\Gamma}.$$

On the other hand, taking (2.29) into account, we can write

$$\langle T\{e^{\Phi_0+\xi(\Phi-\Phi_0)}[(\Phi-\Phi_0)^2 - (f'(\xi))^2]\}\rangle_\Gamma$$

$$= \langle T\{e^{\Phi_0+\xi(\Phi-\Phi_0)}[(\Phi-\Phi_0)^2 - (f'(\xi))^2 - 2f'(\xi)(\Phi-\Phi_0 - f'(\xi))]\}\rangle_\Gamma$$

$$= \langle T\{e^{\Phi_0+\xi(\Phi-\Phi_0)}(\Phi-\Phi_0 - f'(\xi))^2\}\rangle_\Gamma,$$

so that

$$f''(\xi) = \frac{\langle T\{e^{\Phi_0+\xi(\Phi-\Phi_0)}(\Phi-\Phi_0 - f'(\xi))^2\}\rangle_\Gamma}{\langle T\{e^{\Phi_0+\xi(\Phi-\Phi_0)}\}\rangle_\Gamma}. \tag{2.30}$$

We see that the expressions

$$e^{\Phi_0+\xi(\Phi-\Phi_0)}, \quad e^{\Phi_0+\xi(\Phi-\Phi_0)}(\Phi-\Phi_0 - f'(\xi))^2$$

are positive for arbitrary real $\mathbf{R}(s)$ if they are considered to be functionals of $\mathbf{R}(s)$ for $(0 \leqslant s \leqslant \beta\hbar)$. Thus, using the property of the Γ-averages for T-products investigated before (see (1.127)), we can show that

$$f''(\xi) \geqslant 0. \tag{2.31}$$

It follows from (2.31) that

$$f''(\xi) \geqslant f'(0) \quad \text{if} \quad \xi \geqslant 0$$

and

$$f(1) - f(0) = \int_0^1 f'(\xi)\, d\xi \geqslant f'(0),$$

or

$$f(0) + f'(0) \leqslant f(1).$$

Multiplying the last inequality by $-\vartheta$, we obtain

$$-\vartheta f(0) - \vartheta f'(0) \geqslant -\vartheta f(1).$$

Taking (2.17), (2.25), (2.27) and (2.28) into account, we get

$$-\vartheta f(1) = F_{\text{int}}^{(P)}(V, \eta),$$

$$-\vartheta f(0) - \vartheta f'(0) = \text{app } F_{\text{int}}^{(P)}(V, \eta),$$

and consequently our main inequality (2.26) is proved. It can now be noted that, because of (2.20) and (2.25), we can write

$$\text{app } F_{\text{int}}^{(P)}(V, \eta) = \text{app } F_{\text{int}}^{(L)}(V, \eta) - \vartheta \frac{\langle T\{e^{\Phi_0}(\Phi-\Phi_0)\}\rangle_\Gamma}{\langle T\{e^{\Phi_0}\}\rangle_\Gamma}. \tag{2.32}$$

It is worth stressing that the quantity $F_{\text{int}}^{(L)}(V, \eta)$ has been calculated in Chapter 1 (see (1.50)–(1.54)), as well as the limiting relation

$$F_{\text{int}}^{(L)} = \lim_{\eta\to 0} \lim_{V\to\infty} F_{\text{int}}^{(L)}(V, \eta)$$

(see (1.54)). For the special case when $H^{(L)}(\Sigma)$ is a single-frequency Hamiltonian, i.e. when $\nu(f) = \nu = \text{const}$, we obtain a simple closed-form expression (1.56) from the Chapter 1 for $F_{\text{int}}^{(L)}$.

Let us now transform the second term on the right-hand side of (2.32). Taking (2.16) and (2.24) into account, we obtain

$$\frac{\langle T\{e^{\Phi}(\Phi - \Phi_0)\}\rangle_{\Gamma}}{\langle T\{e^{\Phi_0}\}\rangle_{\Gamma}} = \frac{1}{2\hbar^2 V} \sum_{(f)} L^2(f) \frac{\hbar}{2\omega} (1 - e^{-\beta\hbar\omega})^{-1}$$

$$\times \int_0^{\hbar\beta} ds \int_0^{\hbar\beta} d\sigma \, (e^{-\omega|s-\sigma|} + e^{-\omega\beta\hbar + \omega|s-\sigma|}) \frac{\langle T\{e^{\Phi_0 + if\cdot\{\mathbf{R}(s) - \mathbf{R}(\sigma)\}}\}\rangle_{\Gamma}}{\langle T\{e^{\Phi_0}\}\rangle_{\Gamma}}$$

$$+ \int_0^{\hbar\beta} ds \int_0^{\hbar\beta} d\sigma \, L(|s - \sigma|) \frac{\langle T\{e^{\Phi_0}[\mathbf{R}(s) - \mathbf{R}(\sigma)]^2\}\rangle_{\Gamma}}{\langle T\{e^{\Phi_0}\}\rangle_{\Gamma}}.$$

Thus we need only to transform the expressions

$$\frac{\langle T\{e^{\Phi_0 + if\cdot[\mathbf{R}(s) - \mathbf{R}(\sigma)]}\}\rangle_{\Gamma}}{\langle T\{e^{\Phi_0}\}\rangle_{\Gamma}} \tag{2.33}$$

$$\frac{\langle T\{e^{\Phi_0}[\mathbf{R}(s) - \mathbf{R}(\sigma)]^2\}\rangle_{\Gamma}}{\langle T\{e^{\Phi_0}\}\rangle_{\Gamma}}. \tag{2.34}$$

To do this, we first calculate the auxiliary expression

$$\langle T\{e^{\Phi_0 + \int_0^{\hbar\beta} ds\, \lambda(s)\, \mathbf{f}\cdot\mathbf{R}(s)}\}\rangle_{\Gamma}, \tag{2.35}$$

in which we can put for example

$$\lambda(s) = i[\delta(s - s_1) - \delta(s - s_0)], \tag{2.36}$$

$$0 < s_0 < \beta\hbar, \quad 0 < s_1 < \beta\hbar.$$

Consider the operator equation

$$\hbar\frac{dU(s)}{ds} = -[H^{(L)} - \hbar\lambda(s)\, \mathbf{f}\cdot\mathbf{r}]U(s)$$

$$= -[H(S) + H(\Sigma) + H_{\text{int}}^{(L)}(S, \Sigma) - \hbar\lambda(s)\, \mathbf{f}\cdot\mathbf{r}]U(S), \tag{2.37}$$

$$U(0) = 1.$$

This equation corresponds to (1.136), in which we put

$$H_0 = H(S) + H^{(L)}(\Sigma), \quad H_1(S) = H_{\text{int}}^{(L)}(S, \Sigma) - \hbar\lambda(s)\, \mathbf{f}\cdot\mathbf{r}.$$

We repeat our previous reasoning when calculating the magnitude

$$\frac{\text{Tr}_{S,\Sigma}\, U(\beta\hbar)}{\text{Tr}_{(S)}\, e^{-\beta H(S)}\, \text{Tr}_{\Sigma}\, e^{-\beta H^{(L)}(\Sigma)}},$$

noting that in this case, when using (1.135), we have to choose

$$A(S) = \frac{K_0^2 \mathbf{R}^2(S)}{2\hbar} + \lambda(S)\,\mathbf{f}\cdot\mathbf{R}(s).$$

Thus we can find that

$$\frac{\mathrm{Tr}_{S,\Sigma}\,U(\beta\hbar)}{\mathrm{Tr}_{(S)}\,e^{-\beta H(S)}\,\mathrm{Tr}_\Sigma\,e^{-\beta H^{(L)}(\Sigma)}} = \left\langle T\Big\{e^{\Phi_0 + \int\limits_0^{\hbar\beta} ds\lambda(s)\,\mathbf{f}\cdot\mathbf{R}(s)}\Big\}\right\rangle_\Gamma.$$

But, because of (2.20),

$$\frac{\mathrm{Tr}_{S,\Sigma}\,e^{-\beta H^{(L)}}}{\mathrm{Tr}_{(S)}\,e^{-\beta H(S)}\,\mathrm{Tr}_\Sigma\,e^{-\beta H^{(L)}(\Sigma)}} = \langle T\{e^{\Phi_0}\}\rangle_\Gamma.$$

Thus

$$\frac{\left\langle T\Big\{e^{\Phi_0 + \int\limits_0^{\beta\hbar} ds\,\lambda(s)\,\mathbf{f}\cdot\mathbf{R}(s)}\Big\}\right\rangle_\Gamma}{\langle T\{e^{\Phi_0}\}\rangle_\Gamma} = \frac{\mathrm{Tr}_{S,\Sigma}\,U(\beta\hbar)}{\mathrm{Tr}_{S,\Sigma}\,e^{-\beta H^{(L)}}}. \qquad (2.38)$$

Let us now return to (1.136) and (2.37), and choose

$$H_0 = H^{(L)}, \qquad H_1(s) = -\hbar\lambda(s)\,\mathbf{f}\cdot\mathbf{r}.$$

Then, because of (1.137), we find that

$$U(\hbar\beta) = e^{-\beta H^{(L)}} T\left\{\exp\left(\int\limits_0^{\hbar\beta} ds\,\lambda(S)e^{H^{(L)}\frac{s}{\hbar}}\,\mathbf{f}\cdot\mathbf{r}\,e^{-H^{(L)}\frac{s}{\hbar}}\right)\right\}. \qquad (2.39)$$

We note that in the Heisenberg representation the operator $\mathbf{r}(t)$ defined by the linear equation of motion, which is induced by the Hamiltonian $H^{(L)}$, would be as follows:

$$\mathbf{r}(t) = e^{i\frac{t}{\hbar}H^{(L)}}\,\mathbf{r}\,e^{-i\frac{t}{\hbar}H^{(L)}},$$

Therefore

$$e^{\frac{s}{\hbar}H^{(L)}}\,\mathbf{r}\,e^{-\frac{s}{\hbar}H^{(L)}} = \mathbf{r}(-is).$$

It follows from (2.39) that

$$U(\hbar\beta) = e^{-\beta H^{(L)}} T\left\{e^{\int\limits_0^{\hbar\beta} ds\lambda(s)\,\mathbf{f}\cdot\mathbf{r}(-is)}\right\}.$$

Thus

$$\frac{\mathrm{Tr}_{S,\Sigma}\,U(\beta\hbar)}{\mathrm{Tr}_{S,\Sigma}\,e^{-\beta H^{(L)}}} = \left\langle T\Big\{e^{\int\limits_0^{\hbar\beta} ds\lambda(s)\,\mathbf{f}\cdot\mathbf{r}(-is)}\Big\}\right\rangle_{H^{(L)}}. \qquad (2.40)$$

Because $H^{(L)}$ is also a positive-definite form composed of Bose operators, we can use (1.82) and write down

$$\left\langle T\left\{e^{\int_0^{\hbar\beta} ds \lambda(s)\, \mathbf{f}\cdot\mathbf{r}(-is)}\right\}\right\rangle_{H^{(L)}} = \exp\left\langle T\left\{\left(\int_0^{\hbar\beta} ds\, \lambda(s)\,\mathbf{f}\cdot\mathbf{r}(-is)\right)^2\right\}\right\rangle_{H^{(L)}}.$$

But (1.52) gives the identities

$$\langle r_\alpha(t) r_{\alpha'}(\tau)\rangle_{H^{(L)}} = 0 \quad \text{if} \quad \alpha \neq \alpha',$$
$$\langle r_\alpha(t) r_\alpha(\tau)\rangle_{H^{(L)}} = \langle r_1(t) r_1(\tau)\rangle_{H^{(L)}}, \quad \alpha = 1, 2, 3,$$

from which it follows that

$$\left\langle T\left\{\int_0^{\hbar\beta} ds\, \lambda(s) r_\alpha(-is) \int_0^{\hbar\beta} ds\, \lambda(s) r_{\alpha'}(-is)\right\}\right\rangle_{H^{(L)}} = 0, \quad \alpha \neq \alpha',$$

$$\left\langle T\left\{\left(\int_0^{\hbar\beta} ds\, \lambda(s) r_\alpha(-is)\right)^2\right\}\right\rangle_{H^{(L)}} = \left\langle T\left\{\left(\int_0^{\hbar\beta} ds\, \lambda(s) r_1(-is)\right)^2\right\}\right\rangle_{H^{(L)}}$$

and

$$\left\langle T\left\{\left(\int_0^{\hbar\beta} ds\, \lambda(s)\mathbf{f}\cdot\mathbf{r}(-is)\right)^2\right\}\right\rangle_{H^{(L)}} = f^2 \left\langle \frac{1}{2} T\left\{\left(\int_0^{\hbar\beta} ds\, \lambda(s) r_1(-is)\right)^2\right\}\right\rangle_{H^{(L)}}$$

$$= \frac{f^2}{6}\left\langle T\left\{\sum_{\alpha=1}^{3}\left(\int_0^{\hbar\beta} ds\, \lambda(s) r_\alpha(-is)\right)^2\right\}\right\rangle_{H^{(L)}}$$

$$= \frac{f^2}{6}\left\langle T\left\{\left(\int_0^{\hbar\beta} ds\, \lambda(s)\mathbf{r}(-is)\right)^2\right\}\right\rangle_{H^{(L)}}.$$

We can derive from (2.38) and (2.40) that

$$\frac{\left\langle T\left\{e^{\Phi_0} e^{\int_0^{\hbar\beta} ds\lambda(s)\,\mathbf{f}\cdot\mathbf{R}(s)}\right\}\right\rangle_\Gamma}{\langle T\{e^{\Phi_0}\}\rangle_\Gamma} = \exp\frac{f^2}{6}\left\langle T\left\{\left(\int_0^{\hbar\beta} ds\lambda(s)\,\mathbf{r}(-is)\right)^2\right\}\right\rangle_{H^{(L)}}.$$
$$\tag{2.41}$$

In particular, for the function $\lambda(s)$ defined by (2.36), relation (2.41) results in

$$\frac{\langle T\{e^{\Phi_0 + i\mathbf{f}\cdot[\mathbf{R}(s_1)-\mathbf{R}(s_2)]}\}\rangle_\Gamma}{\langle T\{e^{\Phi_0}\}\rangle_\Gamma} = e^{-\frac{f^2}{6}\langle T\{[\mathbf{r}(-is_1)-\mathbf{r}(-is_2)]^2\}\rangle_{H^{(L)}}},$$

or, after a change of variables,

$$\frac{\langle T\{e^{\Phi_0 + i\mathbf{f}\cdot[\mathbf{R}(s)-\mathbf{R}(\sigma)]}\}\rangle_\Gamma}{\langle T\{e^{\Phi_0}\}\rangle_\Gamma} = e^{-\frac{f^2}{6}\langle T\{[\mathbf{r}(-is)-\mathbf{r}(-i\sigma)]^2\}\rangle_{H^{(L)}}}. \tag{2.42}$$

If we apply the operator

$$-\sum_{\alpha=1}^{3} \frac{\partial^2}{\partial f_\alpha^2}$$

to both sides of (2.42) and put $\mathbf{f} = 0$, we obtain

$$\frac{\langle T\{e^{\Phi_0}[\mathbf{R}(s) - \mathbf{R}(\sigma)]^2\}\rangle_\Gamma}{\langle T\{e^{\Phi_0}\}\rangle_\Gamma} = \langle T\{(\mathbf{r}(-is) - \mathbf{r}(-i\sigma))^2\}\rangle_{H^{(L)}}. \qquad (2.43)$$

From (2.23), (2.42) and (2.43), we can derive the identity

$$\text{app } F_{\text{int}}^{(P)}(V, \eta) = \text{app } F_{\text{int}}^{(L)}(V, \eta) - \vartheta \frac{1}{2\hbar^2 V} \sum_{(f)} L^2(f) \frac{\hbar}{2\omega} (1 - e^{-\beta\omega\hbar})^{-1}$$

$$\times \int_0^{\hbar\beta} ds \int_0^{\hbar\beta} d\sigma \, (e^{-\omega|s-\sigma|} + e^{-\omega\beta\hbar+\omega|s-\sigma|}) e^{-\frac{f^2}{6}\langle T\{[\mathbf{r}(-is)-\mathbf{r}(-i\sigma)]^2\}\rangle_{H^{(L)}}}$$

$$- \vartheta \frac{1}{4\hbar^2 V} \sum_{(f)} \frac{S^2(f) f^2 \hbar}{6\nu(f)} (1 - e^{-\beta\hbar\nu(f)})^{-1}$$

$$\times \int_0^{\hbar\beta} ds \int_0^{\hbar\beta} d\sigma \, (e^{-\nu(f)|s-\sigma|} + e^{-\nu(f)\beta\hbar+\nu(f)|s-\sigma|})\langle T\{[\mathbf{r}(-is) - \mathbf{r}(-i\sigma)]^2\}\rangle_{H^{(L)}}.$$

$$(2.44)$$

We should note that the expression

$$\langle T\{[\mathbf{r}(-is) - \mathbf{r}(-i\sigma)]^2\}\rangle_{H^{(L)}}$$

has already been considered in Chapter 1. So, keeping in mind (1.71) and (1.72), we can write

$$\langle T\{[\mathbf{r}(-is) - \mathbf{r}(-i\sigma)]^2\}\rangle_{H^{(L)}}$$

$$= \frac{3i\hbar}{2\pi} \int_{-\infty}^{+\infty} \frac{2(1 - e^{-\nu|s-\sigma|})}{1 - e^{-\hbar\nu\beta}} \frac{1}{m\Omega^2 - \eta\Omega + \Omega\Delta(\Omega)} \Big|_{\nu-i0}^{\nu+i0} d\nu, \quad (2.45)$$

where (see (1.20))

$$\Delta(\Omega) = -\frac{1}{V} \sum_{(f)} \frac{S^2(f)}{6\nu^2(f)} f^2 \left(\frac{1}{\Omega + \nu(f)} + \frac{1}{\Omega - \nu(f)}\right). \qquad (2.46)$$

We are now going to consider the classical limit of the expression (2.44). To make a transition to classical mechanics, we should take the limit $\hbar \to 0$. It is convenient to introduce new variables in (2.44):

$$s = \hbar s', \quad \sigma = \hbar\sigma'.$$

Then we can see from (2.44) and (2.45) that

$$\text{app } F_{\text{int}}^{(P)}(V, \eta) = \text{app } F_{\text{int}}^{(L)}(V, \eta) - \frac{\vartheta}{2V} \sum_{(f)} L^2(f) \frac{\hbar}{2\omega(1 - e^{-\beta\hbar\omega})}$$

$$\times \int_0^\beta ds' \int_0^\beta d\sigma' \frac{(e^{-\omega\hbar|s'-\sigma'|} + e^{-\omega\hbar\beta+\omega\hbar|s'-\sigma'|})}{e^{\frac{f^2}{6}\langle T\{[\mathbf{r}(-i\hbar s')-\mathbf{r}(-i\hbar\sigma')]^2\}\rangle_{H^{(L)}}}}$$

$$-\frac{\vartheta}{4V}\sum_{(f)}\frac{S^2(f)f^2}{6\nu(f)}\frac{\hbar}{2\omega(1-e^{-\beta\hbar\nu(f)})}$$

$$\times\int_0^\beta ds'\int_0^\beta d\sigma'\frac{(e^{-\nu(f)\hbar|s'-\sigma'|}+e^{-\nu(f)\hbar\beta+\nu(f)\hbar|s'-\sigma'|})}{e^{\frac{f^2}{6}\langle T\{[\mathbf{r}(-i\hbar s')-\mathbf{r}(-i\hbar\sigma')]^2\}\rangle_{H(L)}}},\qquad(2.47)$$

$$\langle T\{[\mathbf{r}(-i\hbar s')-\mathbf{r}(-i\hbar\sigma')]^2\}\rangle_{H(L)}$$

$$=\frac{3i}{2\pi}\int_{-\infty}^{+\infty}\frac{2\hbar}{1-e^{-\hbar\nu\beta}}(1-e^{-\nu|s'-\sigma'|\hbar})\frac{1}{m\Omega^2-\eta\Omega+\Omega\Delta(\Omega)}\bigg|_{\nu-i0}^{\nu+i0}d\nu,\qquad(2.48)$$

If the variables V and η are fixed then the expression for $\Delta(\Omega)$ contains only a finite number of terms. Hence we see that

$$\langle T\{[\mathbf{r}(-i\hbar s')-\mathbf{r}(-i\hbar\sigma')]^2\}\rangle_{H(L)}\to 0\quad\text{for}\quad\hbar\to 0.$$

Observing that all sums in (2.47) contain only a finite number of terms, we have in the classical limit

$$\text{app}\,F_{\text{int}}^{(P)}(V,\eta)=-\frac{1}{2V\omega^2}\sum_{(f)}L^2(f),\qquad(2.49)$$

because it was shown earlier that for classical mechanics the following identity holds

$$\text{app}\,F_{\text{int}}^{(L)}(V,\eta)=0.$$

We shall show now that (2.49) gives the exact value of the free energy app $F_{\text{int}}^{(P)}$ in the classical limit. In fact, the Hamiltonian $H^{(P)}$ is the sum of the potential and kinetic energies. Therefore, in the case of classical mechanics,

$$F^{(P)}=\text{kin}\,F^{(P)}+\text{pot}\,F^{(P)},\qquad \text{kin}\,F^{(P)}=\text{kin}\,F_S+\text{kin}\,F_\Sigma$$

and

$$F_{\text{int}}^{(P)}=F^{(P)}-F_S-F_\Sigma=\text{pot}\,F^{(P)}-\text{pot}\,F_S-\text{pot}\,F_\Sigma.$$

Therefore we arrive at the expression

$$F_{\text{int}}^{(P)}(V,\eta)$$

$$=-\vartheta\ln\frac{\int\exp\left(-\beta\frac{\eta^2\mathbf{r}^2}{2}-\frac{\omega^2}{2}\beta\sum_{(f)}q_fq_f^\dagger-\frac{\beta}{V^{1/2}}\sum_{(f)}L(f)q_fe^{i\mathbf{f}\cdot\mathbf{r}}\right)d\mathbf{r}\,Dq}{\int\exp\left(-\beta\frac{\eta^2\mathbf{r}^2}{2}\right)d\mathbf{r}\int\exp\left(-\frac{\beta\omega^2}{2}\sum_{(f)}q_fq_f^\dagger\right)Dq}.$$

$$(2.50)$$

Here
$$q_f = x_f + iy_f, \quad q_{-f} = x_f - iy_f,$$

i.e.
$$x_{-f} = x_f, \quad y_{-f} = -y_f,$$

and
$$Dq = \prod_{(f)}{}' dx_f \, dy_f.$$

Here the symbol $\prod'$ denotes that only different dx_f and dy_f are included; thus if $f \neq 0$ is included in the product then $-f$ is not included. After the change of variables

$$q_f \rightarrow q_f - \frac{L(f)}{\omega^2 V^{1/2}} e^{-i\mathbf{f} \cdot \mathbf{r}}, \qquad q_f^\dagger \rightarrow q_f^\dagger - \frac{L(f)}{\omega^2 V^{1/2}} e^{i\mathbf{f} \cdot \mathbf{r}},$$

the expression

$$\frac{\omega^2}{2} \beta \sum_{(f)} q_f q_f^\dagger + \frac{\beta}{V^{1/2}} \sum_{(f)} L(f) q_f e^{i\mathbf{f} \cdot \mathbf{r}}$$

transforms into

$$\frac{\omega^2}{2} \beta \sum_{(f)} q_f q_f^\dagger - \frac{\omega^2}{2} \beta \sum_{(f)} \frac{L(f)}{V^{1/2}} \frac{e^{-i\mathbf{f} \cdot \mathbf{r}} q_f^\dagger + e^{i\mathbf{f} \cdot \mathbf{r}} q_f}{\omega^2}$$

$$+ \frac{\omega^2 \beta}{2} \sum_{(f)} \frac{L^2(f)}{\omega^4 V} + \frac{\beta}{V^{1/2}} \sum_{(f)} L(f) q_f e^{i\mathbf{f} \cdot \mathbf{r}} - \frac{\beta}{V} \sum_{(f)} \frac{L^2(f)}{\omega^2}$$

$$= \frac{\omega^2}{2} \sum_{(f)} q_f q_f^\dagger - \frac{\beta}{2\omega^2 V} \sum_{(f)} L^2(f).$$

Therefore (2.50) can be represented as

$$F_{\text{int}}^{(P)}(V, \eta)$$

$$= -\vartheta \ln \frac{\int \exp\left(-\beta \frac{\eta^2 \mathbf{r}^2}{2} - \frac{\omega^2}{2} \beta \sum_{(f)} q_f q_f^\dagger + \frac{\beta}{2\omega^2 V} \sum_{(f)} L^2(f) \right) d\mathbf{r} \, Dq}{\int \exp\left(-\beta \frac{\eta^2 \mathbf{r}^2}{2} \right) d\mathbf{r} \int \exp\left(-\frac{\beta \omega^2}{2} \sum_{(f)} q_f q_f^\dagger \right) Dq}$$

$$= -\frac{1}{2V\omega^2} \sum_{(f)} L^2(f). \quad (2.51)$$

Hence we see that in classical mechanics the approximation given by (2.49) leads to the exact result.

Now we shall investigate (2.44) in the quantum case. It is convenient to use the function

$$E_V(\omega) = \frac{1}{V} \sum_{(f)} \frac{S^2(f)}{6\nu^2(f)} \, f^2 [\delta(\nu(f) + \omega) + \delta(\nu(f) - \omega)] \tag{2.52}$$

introduced in Chapter 1. With this function, (2.44) can be rewritten in the form

$$\text{app } F_{\text{int}}^{(P)}(V, \eta) = F_{\text{int}}^{(L)}(V, \eta)$$

$$- \frac{\vartheta}{4V\hbar^2} \sum_{(f)} L^2(f) \frac{\hbar}{\omega(1 - e^{-\beta\hbar\omega})} \int_0^{\hbar\beta} ds \int_0^{\hbar\beta} d\sigma \frac{(e^{-\omega|s-\sigma|} + e^{-\beta\hbar\omega+\omega|s-\sigma|})}{e^{\frac{f^2}{6} D_V(s,\sigma)}}$$

$$- \frac{\vartheta}{4\hbar^2} \int_0^\infty d\omega \, E_V(\omega) \frac{\hbar\omega}{1 - e^{-\beta\hbar\omega}} \int_0^{\hbar\beta} ds \int_0^{\hbar\beta} d\sigma (e^{-\omega|s-\sigma|} + e^{-\beta\hbar\omega+\omega|s-\sigma|}) D_V(s,\sigma), \tag{2.53}$$

where (see (2.45), (2.46) and (2.52))

$$D_V(s,\sigma) = \frac{3i\hbar}{\pi} \int_{-\infty}^{+\infty} \frac{1 - e^{-\omega|s-\sigma|}}{1 - e^{-\hbar\omega\beta}} \frac{1}{m\Omega^2 - \eta^2 + \Omega\Delta(\Omega)} \Big|_{\omega-i0}^{\omega+i0} d\omega, \tag{2.54}$$

$$\Delta(\Omega) = - \int_{-\infty}^{+\infty} \frac{E_V(\omega)}{\Omega - \omega}. \tag{2.55}$$

We note that, because of (2.52),

$$E_V(\omega) \geqslant 0, \qquad E_V(-\omega) = E_V(\omega). \tag{2.56}$$

It follows from these relations and from (2.55) that

$$\Delta(-\Omega) = - \int_{-\infty}^{+\infty} \frac{E(\omega)}{-\Omega - \omega} = - \int_{-\infty}^{+\infty} \frac{E(\omega)}{-\Omega + \omega} = -\Delta(\Omega). \tag{2.57}$$

Let us introduce new notation for brevity:

$$\frac{1}{m\Omega^2 - \eta^2 + \Omega\Delta(\Omega)} = \Phi(\Omega), \qquad \frac{3i\hbar}{\pi} \left(\frac{1 - e^{-\omega|s-\sigma|}}{1 - e^{-\hbar\omega\beta}} \right) = \phi(\omega).$$

Then the right-hand side of (2.54) can be expressed as

$$\int_{-\infty}^{+\infty} \phi(\omega) \{ \Phi(\omega + i0) - \Phi(\omega - i0) \}.$$

But it follows from (2.57) that

$$\Phi(\Omega) = \Phi(-\Omega),$$

and in particular,

$$\Phi(\omega + i0) - \Phi(\omega - i0) = \Phi(-\omega - i0) - \Phi(-\omega + i0).$$

From here

$$\int\limits_{-\infty}^{+\infty}\phi(\omega)\{\Phi(\omega+i0)-\Phi(\omega-i0)\}d\omega = \int\limits_{-\infty}^{+\infty}\phi(-\omega)\{\Phi(\omega-i0)-\Phi(\omega+i0)\}d\omega,$$

that yields

$$\int\limits_{-\infty}^{+\infty}\phi(\omega)\{\Phi(\omega+i0)-\Phi(\omega-i0)\}\,d\omega$$

$$=\frac{1}{2}\int\limits_{-\infty}^{+\infty}\{\phi(\omega)-\phi(-\omega)\}\{\Phi(\omega+i0)-\Phi(\omega-i0)\}\,d\omega.$$

On the other hand,

$$\frac{1-e^{-\omega|s-\sigma|}}{1-e^{-\hbar\omega\beta}}-\frac{1-e^{\omega|s-\sigma|}}{1-e^{\hbar\omega\beta}}=\frac{1+e^{-\hbar\omega\beta}-e^{-\omega|s-\sigma|}-e^{-\beta\hbar\omega+\omega|s-\sigma|}}{1-e^{-\hbar\omega\beta}}.$$

Therefore we can derive from (2.54)

$$D_V(s,\sigma)=\frac{3i\hbar}{2\pi}\int\limits_{-\infty}^{+\infty}\frac{1+e^{-\hbar\omega\beta}-e^{-\omega|s-\sigma|}-e^{-\beta\hbar\omega+\omega|s-\sigma|}}{(1-e^{-\hbar\omega\beta})(m\Omega^2-\eta^2+\Omega\Delta(\Omega))}\bigg|_{\omega-i0}^{\omega+i0}d\omega.\quad(2.58)$$

It will be noted that (1.91) and (1.102) give us the following result:

$$\hbar\frac{e^{-\omega|s-\sigma|}+e^{-\hbar\omega\beta+\omega|s-\sigma|}}{2\omega(1-e^{-\hbar\omega\beta})}=\frac{1}{\beta}\sum\limits_{(n)}\frac{e^{in2\pi\frac{s-\sigma}{\beta\hbar}}}{(2\pi n/\beta\hbar)^2+\omega^2}$$

$$=\frac{1}{\beta}\sum\limits_{(n)}\frac{\cos\{2\pi n(s-\sigma)/\beta\hbar\}}{(2\pi n/\beta\hbar)^2+\omega^2}.\quad(2.59)$$

Thus

$$\frac{\hbar(1+e^{-\hbar\omega\beta}-e^{-\omega|s-\sigma|}-e^{-\beta\hbar\omega+\omega|s-\sigma|})}{2\omega(1-e^{-\hbar\omega\beta})}=\frac{1}{\beta}\sum\limits_{(n\neq0)}\frac{1-e^{in2\pi\frac{s-\sigma}{\beta\hbar}}}{(2\pi n/\beta\hbar)^2+\omega^2}$$

$$=\frac{1}{\beta}\sum\limits_{(n\neq0)}\frac{1-\cos\{2\pi n(s-\sigma)/\beta\hbar\}}{(2\pi n/\beta\hbar)^2+\omega^2}.\quad(2.60)$$

These expressions are valid in the domain

$$0\leqslant s\leqslant\beta\hbar,\qquad0\leqslant\sigma\leqslant\beta\hbar,\quad(2.61)$$

but their right-hand sides are defined for all real s and σ and they are periodic functions of $s-\sigma$ with period $\beta\hbar$. It follows from (2.58) that

$$D_V(s,\sigma)=D_V(s-\sigma),\quad(2.62)$$

where

$$D_V(s) = \frac{3i\hbar}{2\pi} \int\limits_{-\infty}^{+\infty} \frac{1 + e^{-\hbar\omega\beta} - e^{-\omega|s|} - e^{-\beta\hbar\omega|s|}}{(1 - e^{-\hbar\omega\beta})(m\Omega^2 - \eta^2 + \Omega\Delta(\Omega))} \bigg|_{\omega-i0}^{\omega+i0} d\omega, \qquad (2.63)$$

This function can be continued to the whole real axis,

$$D_V(s) = \frac{3i}{\pi\beta} \sum_{n\neq 0} \int\limits_{-\infty}^{+\infty} \frac{1 - \cos\{2\pi ns/\beta\hbar\}}{(2\pi n/\beta\hbar)^2 + \omega^2} \frac{\omega}{m\Omega^2 - \eta^2 + \Omega\Delta(\Omega)} \bigg|_{\omega-i0}^{\omega+i0} d\omega, \qquad (2.64)$$

such that it is periodic and symmetric one:

$$D_V(-s) = D_V(s), \qquad D_V(s + \beta\hbar) = D_V(s). \qquad (2.65)$$

Let us consider the functions

$$\mathcal{R}^{(P)}(s) = (e^{-\omega s} + e^{-\beta\hbar\omega+\omega s})e^{-\frac{f^2}{6}D_V(s)}, \quad 0 \leqslant s \leqslant \beta\hbar,$$

$$\mathcal{R}^{(L)}(s) = (e^{-\omega s} + e^{-\beta\hbar\omega+\omega s})D_V(s).$$

Taking (2.59) and (2.64) into account, we can show that these functions can be continued to the whole real axis s in such a way that they possess properties of periodicity and symmetry similar to those given by (2.65). Hence we have the identities

$$\mathcal{R}^{(P,L)}(s) = \sum_{(n)} \mathcal{R}_n^{(P,L)} e^{2\pi in\frac{s}{\beta\hbar}}$$

and

$$\mathcal{R}^{(P,L)}(s - \sigma) = \sum_{(n)} \mathcal{R}_n^{(P,L)} e^{2\pi in\frac{s-\sigma}{\beta\hbar}},$$

from which it follows that

$$\int\limits_0^{\hbar\beta} d\sigma\, \mathcal{R}^{(P,L)}(s - \sigma) = \mathcal{R}_0^{(P,L)} \beta\hbar = \int\limits_0^{\hbar\beta} d\sigma\, \mathcal{R}^{(P,L)}(\sigma),$$

$$\int\limits_0^{\hbar\beta} ds \int\limits_0^{\hbar\beta} d\sigma\, \mathcal{R}^{(P,L)}(s - \sigma) = \beta\hbar \int\limits_0^{\hbar\beta} d\sigma\, \mathcal{R}^{(P,L)}(\sigma). \qquad (2.66)$$

Thanks to (2.65), which are satisfied by the functions $\mathcal{R}^{(P,L)}(s)$ too, we have

$$\int\limits_0^{\hbar\beta} d\sigma\, \mathcal{R}^{(P,L)}(\sigma) = \int\limits_0^{\beta\hbar/2} d\sigma\, \mathcal{R}^{(P,L)}(\sigma) + \int\limits_{\beta\hbar/2}^{\beta\hbar} d\sigma\, \mathcal{R}^{(P,L)}(\sigma)$$

$$= \int\limits_0^{\beta\hbar/2} d\sigma\, \mathcal{R}^{(P,L)}(\sigma) + \int\limits_0^{\beta\hbar/2} d\sigma\, \mathcal{R}^{(P,L)}(\beta\hbar - \sigma) = 2\int\limits_0^{\beta\hbar/2} d\sigma\, \mathcal{R}^{(P,L)}(\sigma). \qquad (2.67)$$

Therefore (2.53) can be transformed into the following:

$$\text{app } F_{\text{int}}^{(P)}(V,\eta) = \text{app } F_{\text{int}}^{(L)}(V,\eta)$$

$$- \frac{1}{4V} \sum_{(f)} L^2(f) \frac{1}{\omega(1-e^{-\beta\omega\hbar})} \int_0^{\hbar\beta} d\sigma \, (e^{-\omega\sigma} + e^{-\beta\hbar\omega+\omega\sigma}) e^{\frac{f^2}{6} D_V(\sigma)}$$

$$- \frac{1}{4} \int_0^{\infty} d\omega \, E_V(\omega) \frac{\omega}{1-e^{-\beta\omega\hbar}} \int_0^{\hbar\beta} d\sigma \, (e^{-\omega\sigma} + e^{-\beta\hbar\omega+\omega\sigma}) D_V(\sigma), \quad (2.68)$$

or, equivalently,

$$\text{app } F_{\text{int}}^{(P)}(V,\eta) = \text{app } F_{\text{int}}^{(L)}(V,\eta)$$

$$- \frac{1}{2V} \sum_{(f)} L^2(f) \frac{1}{\omega(1-e^{-\beta\omega\hbar})} \int_0^{\beta\hbar/2} d\sigma \, (e^{-\omega\sigma} + e^{-\beta\hbar\omega+\omega\sigma}) e^{-\frac{f^2}{6} D_V(\sigma)}$$

$$- \frac{1}{2} \int_0^{\infty} d\omega \, E_V(\omega) \frac{\omega}{1-e^{-\beta\omega\hbar}} \int_0^{\beta\hbar/2} d\sigma \, (e^{-\omega\sigma} + e^{-\beta\hbar\omega+\omega\sigma}) D_V(\sigma). \quad (2.69)$$

Let us proceed with the passage to the limit

$$\lim_{\eta\to 0} \lim_{V\to\infty} . \quad (2.70)$$

Then, using the identity (2.26), we have

$$\text{app } F_{\text{int}}^{(P)} \geqslant F_{\text{int}}^{(P)}, \quad (2.71)$$

where

$$\text{app } F_{\text{int}}^{(P)} = \lim_{\eta\to 0} \lim_{V\to\infty} \text{app } F_{\text{int}}^{(P)}(V,\eta). \quad (2.72)$$

It should be stressed that the functions $S(f)$ and $\nu(f)$, which characterize the auxiliary Hamiltonian $H^{(L)}$, could be chosen arbitrarily.

To calculate (2.72), it is convenient to choose for $S(f)$ and $\nu(f)$ some continuous functions on the real axis, such that

$$\nu(f) > 0, \quad \frac{1}{V} \sum_{|f|>R} \frac{S(f)}{\nu^2(f)} f^2 \leqslant \delta\left(\frac{1}{R}\right) \underset{R\to\infty}{\to} 0,$$

where $\delta(1/R)$ does not depend on V. In this case, if $\phi(\omega)$ is an arbitrary continuous and finite function on the real axis then we have

$$\int_{-\infty}^{+\infty} d\omega \, \phi(\omega) E_V(\omega) = \frac{1}{6V} \sum_{(f)} \frac{S^2(f)}{\nu^2(f)} f^2 \{\phi(\nu(f)) + \phi(-\nu(f))\}$$

$$\to \frac{1}{(2\pi)^3} \int \frac{S^2(f)}{6\nu^2(f)} f^2 \{\phi(\nu(f)) + \phi(-\nu(f))\} \, d\mathbf{f}$$

$$= \frac{4\pi}{(2\pi)^3} \int\limits_0^\infty df \, \frac{S^2(f)}{6\nu^2(f)} \, f^4 \{\phi(\nu(f)) + \phi(-\nu(f))\} \quad \text{when} \quad V \to \infty. \quad (2.73)$$

Thus the generalized limit

$$E(\omega) = \lim_{V \to \infty} E_V(\omega) \qquad (2.74)$$

exists and we can write

$$\int\limits_{-\infty}^{+\infty} d\omega \, \phi(\omega) E_V(\omega) \xrightarrow{V \to \infty} \int\limits_{-\infty}^{+\infty} d\omega \, \phi(\omega) E_V(\omega). \qquad (2.75)$$

This function $E(\omega)$ has the following obvious properties:

$$E(\omega) \geqslant 0, \quad E(-\omega) = E(\omega), \qquad (2.76)$$

$$\int\limits_{-\infty}^{+\infty} d\omega \, E(\omega) = \frac{1}{6\pi^2} \int\limits_0^\infty df \, \frac{S^2(f)}{\nu^2(f)} \, f^4.$$

It follows from (2.75), in particular, that

$$\Delta(\Omega) \to \Delta_\infty(\Omega) = - \int\limits_{-\infty}^{+\infty} \frac{E(\omega)}{\Omega - \omega} \, d\omega \quad \text{when} \quad \text{Im}\,\Omega \neq 0, \qquad (2.77)$$

and

$$\Delta_\infty(\Omega) = -\Delta_\infty(-\Omega).$$

Thus, as we can see,

$$E(\omega) = \frac{1}{2\pi i} \{\Delta_\infty(\omega + i0) - \Delta_\infty(\omega - i0)\}$$

$$= \frac{1}{2\pi i} \{\Delta_\infty(\omega + i0) + \Delta_\infty(-\omega + i0)\}. \qquad (2.78)$$

The calculation of (2.72) should be commenced with the proper choice of the function $\Delta_\infty(\Omega)$. Let us choose an analytic function $\Delta^+(\Omega)$ that is regular in the upper half–plane and possesses the following properties:
(1) for large enough R,

$$|\Delta^+(\Omega)| \leqslant \frac{C}{|\Omega|}, \quad C = \text{const}, \quad |\Omega| \geqslant R, \quad \text{Im}\,\Omega \geqslant 0;$$

(2) the generalized function

$$\lim_{\varepsilon \to 0, \varepsilon > 0} \Delta^+(\omega + i\varepsilon) = \Delta^+(\omega + i0)$$

exists on the real axis ω, and the expression

$$\frac{1}{2\pi i} \{\Delta^+(\omega + i0) + \Delta^+(-\omega + i0)\} \geqslant 0.$$

The integral of this expression, taken over the whole real axis, converges. Let us now choose

$$\Delta(\Omega) = \Delta_+(\Omega), \quad \operatorname{Im}\Omega > 0,$$
$$\Delta(\Omega) = -\Delta_+(-\Omega), \quad \operatorname{Im}\Omega > 0. \tag{2.79}$$

Bearing the conditions (1) and (2) in mind, it is easy to see that this function (2.79) and the corresponding function $E(\omega)$ satisfy all the conditions mentioned above.

Supposing Δ_∞ to be fixed, we choose $S(f)$ and $\nu(f)$ in such a way that the limit conditions (2.74)–(2.76) are obliged. To analyze the expression

$$\lim_{\eta \to 0} \lim_{V \to \infty} D_V(s),$$

we start from the definition (2.63), noting that the expression

$$\mathcal{L}(s, \Omega) = \frac{1 + e^{-\hbar\beta\Omega} - e^{-\Omega s} - e^{-\beta\hbar\Omega + \Omega s}}{\Omega(1 - e^{-\beta\hbar\Omega})}, \quad 0 \leqslant s \leqslant \beta\hbar, \tag{2.80}$$

is an analytic function of Ω on the whole complex plane that has poles at

$$\Omega = \frac{2\pi n i}{\beta\hbar}, \quad \text{where } n \text{ is an integer,}$$

and is free from any other singularities. We are going to show now that the point $\Omega = 0$ is not a pole, so the function (2.80) is regular in the vicinity of this point. In fact, the expansion in powers of $\Omega = 0$ results in the series

$$1 + e^{-\hbar\beta\Omega} - e^{-\Omega s} - e^{-\beta\hbar\Omega + \Omega s} = \Omega^2(\beta\hbar s - s^2) + \Omega^3 \dots,$$
$$\Omega(1 - e^{-\beta\hbar\Omega}) = \beta\hbar\Omega^2 + \Omega^3 \dots.$$

These expansions lead to the final expansion for the function in question:

$$\mathcal{L}(s, \Omega) = s\left(1 - \frac{s}{\beta\hbar}\right) + \Omega \dots \tag{2.81}$$

Therefore the only singularities of $\mathcal{L}(s, \Omega)$, as a function of Ω, are simple poles:

$$\Omega = \frac{2\pi n i}{\beta\hbar}, \quad n = \pm 1, \pm 2, \pm 3, \dots$$

These features allows us to rewrite (2.63) and (2.80) in the form

$$D_V(s) = \frac{3i\hbar}{2\pi} \int_{-\infty}^{+\infty} \mathcal{L}(s, \Omega) \frac{\Omega}{m\Omega^2 - \eta^2 + \Omega\Delta(\Omega)} \bigg|_{\omega - i0}^{\omega + i0} d\omega, \tag{2.82}$$

$$0 \leqslant s \leqslant \beta\hbar.$$

This expression can be treated by the method developed in Chapter 1.
Note that the expression

$$\mathcal{L}(s, \Omega) \frac{\Omega}{m\Omega^2 - \eta^2 + \Omega\Delta(\Omega)}, \qquad 0 \leqslant s \leqslant \beta\hbar$$

is nothing other than an analytic function of Ω that is regular on the strips

$$0 < \operatorname{Im}\Omega < \frac{2\pi}{\beta\hbar}, \qquad 0 > \operatorname{Im}\Omega > -\frac{2\pi}{\beta\hbar},$$

and has an order $1/|\Omega|^2$ as $\Omega \to \infty$. Therefore it follows from (2.82) that

$$D_V(s) = \frac{3i\hbar}{2\pi} \int_{-\infty}^{+\infty} \mathcal{L}(s, \Omega) \frac{\Omega}{m\Omega^2 - \eta^2 + \Omega\Delta(\Omega)} \Big|_{\omega - i\varepsilon}^{\omega + i\varepsilon} d\omega, \qquad (2.83)$$

where ε is an arbitrary number from the interval

$$0 < \varepsilon < \frac{2\pi}{\beta\hbar}, \qquad (2.84)$$

and (2.83) does not depend on ε in this interval. Fixing ε , we can go to
the limit

$$\lim_{\eta \to 0} \lim_{V \to \infty}$$

in (2.83), thus obtaining

$$\lim_{\eta \to 0} \lim_{V \to \infty} D_V(s) = \frac{3i\hbar}{2\pi} \int_{-\infty}^{+\infty} \mathcal{L}(s, \Omega) \frac{1}{m\Omega + \Delta_\infty(\Omega)} \Big|_{\omega - i0}^{\omega + i0} d\omega$$

$$= \frac{3i\hbar}{2\pi} \int_{-\infty}^{+\infty} \mathcal{L}(s, \omega) \frac{1}{m\Omega + \Delta_\infty(\Omega)} \Big|_{\omega - i0}^{\omega + i0} d\omega. \qquad (2.85)$$

In order to get the limiting expression

$$\operatorname{app} F_{\text{int}}^{(P)} = \lim_{\eta \to 0} \lim_{V \to \infty} \operatorname{app} F_{\text{int}}^{(P)}(V, \eta)$$

consider the case of the true Fröhlich model with $L(f)$ given by (2.3). In
this case

$$\lim_{\eta \to 0, V \to \infty} \frac{1}{V} \sum_{(f)} L^2(f) e^{-\frac{f^2}{6} D_V(s)} = \frac{1}{(2\pi)^3} \int \frac{g^2}{|f^2|} e^{-\frac{f^2}{6} D_V(s)} d\mathbf{f}$$

$$= \frac{4\pi g^2}{(2\pi)^3} \int_0^\infty e^{-\frac{f^2}{6} D_V(s)} df = \frac{4\pi g^2}{(2\pi)^3} \frac{\pi^{1/2}}{2} \left(\frac{6}{D(s)}\right)^{1/2}. \qquad (2.86)$$

From (2.68) and (2.69), we obtain

$$\text{app } F_{\text{int}}^{(P)} = \text{app } F_{\text{int}}^{(L)} - \frac{g^2\pi}{(2\pi)^3} \frac{\pi^{1/2}}{2} \frac{1}{\omega(1 - e^{-\beta\omega\hbar})}$$

$$\times \int_0^{\hbar\beta} d\sigma \, (e^{-\omega\sigma} + e^{-\omega\hbar\beta+\omega\sigma}) \left(\frac{6}{D(\sigma)}\right)^{1/2} - \frac{1}{4}\int_0^\infty d\omega \, E(\omega)\frac{\omega}{1 - e^{-\beta\omega\hbar}}$$

$$\times \int_0^{\hbar\beta} d\sigma \, (e^{-\omega\sigma} + e^{-\omega\hbar\beta+\omega\sigma})D(\sigma), \quad (2.87)$$

or

$$\text{app } F_{\text{int}}^{(P)} = \text{app } F_{\text{int}}^{(L)}$$

$$- \frac{g^2\pi^{3/2}}{(2\pi)^3} \frac{1}{\omega(1 - e^{-\beta\omega\hbar})} \int_0^{\beta\hbar/2} d\sigma \, (e^{-\omega\sigma} + e^{-\omega\hbar\beta+\omega\sigma}) \left(\frac{6}{D(\sigma)}\right)^{1/2}$$

$$- \frac{1}{2}\int_0^\infty d\omega \, E(\omega)\frac{\omega}{1 - e^{\beta\omega\hbar}} \int_0^{\beta\hbar/2} d\sigma \, (e^{-\omega\sigma} + e^{-\omega\hbar\beta+\omega\sigma})D(\sigma). \quad (2.88)$$

This equation gives a function that satisfies the inequality

$$\text{app } F_{\text{int}}^{(P)} \geqslant F_{\text{int}}^{(P)},$$

so that we can try to construct an effective approximation, considering trial functions $\Delta_\infty(\Omega)$ that contain some variational parameters. These parameters will be fixed later by the minimization condition for $\text{app } F_{\text{int}}^{(P)}(V, \eta)$. To illustrate the idea, we begin with the simplest but meaningful choice

$$\Delta_\infty(\Omega) = -\frac{K_0^2\Omega}{\Omega^2 - \nu_0^2}, \qquad E(\omega) = \frac{K_0^2}{2}\{\delta(\omega - \nu_0) + \delta(\omega + \nu_0)\}. \quad (2.89)$$

As pointed out in Chapter 1, this choice is equivalent to Feynman's model, in which $H^{(L)}$ is the two-body Hamiltonian describing a particle S that interacts with another particle Σ via harmonic force. [h] From (2.89), we have

$$\frac{1}{m\Omega + \Delta_\infty(\Omega)} = \frac{\nu_0^2 - \Omega^2}{m\Omega(\nu_0^2 - \Omega^2)} = \frac{\nu_0^2 - \Omega^2}{m\Omega} \frac{1}{2\mu_0}\left(\frac{1}{\Omega + \mu_0} - \frac{1}{\Omega - \mu_0}\right),$$
$$(2.90)$$

where

$$\mu_0^2 = \frac{K_0^2}{m} + \nu_0^2, \qquad K_0^2 = m(\mu_0^2 - \nu_0^2).$$

---

[h] See also Chapter 8, Section 4 in Feynman's book [5].

Hence

$$\frac{1}{m\Omega + \Delta_\infty(\Omega)} \Big|_{\omega-i0}^{\omega+i0}$$

$$= -2\pi i \left\{ \frac{1}{m} \left(\frac{\nu_0}{\mu_0} \right)^2 \delta(\omega) + \frac{\mu_0^2 - \nu_0^2}{2m\mu_0^2} [\delta(\omega - \mu_0) + \delta(\omega + \mu_0)] \right\}.$$

Now we use (2.85), keeping in mind that, because of (2.80),

$$\mathcal{L}(s,\omega) = \mathcal{L}(s,-\omega).$$

From this, we have

$$D(s) = \frac{3\hbar}{m} \left(\frac{\nu_0}{\mu_0} \right)^2 \mathcal{L}(s,0) + \frac{3\hbar}{m} \frac{\mu_0^2 - \nu_0^2}{\mu_0^2} \mathcal{L}(s,\mu_0).$$

This result can be transformed thanks to (2.80) and (2.81):

$$D(s) = \frac{3\hbar}{m} \left(\frac{\nu_0}{\mu_0} \right)^2 s \left(1 - \frac{s}{\beta\hbar} \right)$$

$$+ \frac{3\hbar}{m} \frac{\mu_0^2 - \nu_0^2}{\mu_0^3} \left(\frac{1 + e^{-\hbar\beta\mu_0} - e^{-s\mu_0} - e^{-\beta\hbar\mu_0+s\mu_0}}{1 - e^{-\beta\hbar\mu_0}} \right), \quad (2.91)$$

where $0 \leqslant s \leqslant \beta\hbar$. This expression has already been derived in Chapter:1 when we considered Feynman's two-body model. Let us note that for the choice (2.89) the corresponding expression for app $F_{\text{int}}^{(L)}$ has also been calculated there:

$$\text{app } F_{\text{int}}^{(L)} = -3\vartheta \ln \frac{\mu_0}{\nu_0} + \frac{3\hbar}{2} (\mu_0 - \nu_0) - 3\vartheta \ln \frac{1 - e^{-\beta\hbar\nu_0}}{1 - e^{-\beta\hbar\mu_0}}. \quad (2.92)$$

Let us substitute now (2.91) and (2.92) into (2.88). It is convenient to introduce dimensionless parameters

$$\mu = \frac{\mu_0}{\omega}, \quad \nu = \frac{\nu_0}{\omega} \quad (2.93)$$

and dimensionless constants

$$\beta_d = \beta\hbar\omega, \quad \alpha = \frac{g^2}{4\pi\hbar\omega^2} \left(\frac{m}{2\hbar\omega} \right)^{1/2}. \quad (2.94)$$

Then, putting

$$z = \omega\sigma, \quad d\sigma = \frac{1}{\omega} dz, \quad 0 \leqslant z \leqslant \beta_d,$$

in the integrand, we we come to the main result of this chapter

$$\frac{\text{app } F_{\text{int}}^{(P)}}{\hbar\omega} = -\frac{3}{\beta_d} \ln \frac{\mu}{\nu} - \frac{3}{\beta_d} \ln \frac{1 - e^{-\beta_d\nu}}{1 - e^{-\beta_d\mu}} + \frac{3}{2} (\mu - \nu)$$

$$- \frac{\alpha}{(1 - e^{-\beta_d})\pi^{1/2}} \int\limits_0^{\beta_d/2} dz \, (e^{-z} + e^{-\beta_d+z})$$

$$\times \left\{ \left(\frac{\nu}{\mu}\right)^2 z \left(1 - \frac{z}{\beta_d}\right) + \frac{\mu^2 - \nu^2}{\mu^3} \frac{1 + e^{-\beta_d\mu} - e^{-\mu z} - e^{-\beta_d\mu + z\mu}}{1 - e^{-\beta_d\mu}} \right\}^{-1/2}$$

$$- \frac{3}{4} \nu(\mu^2 - \nu^2)(1 - e^{-\beta_d\nu})^{-1} \int_0^{\beta_d/2} dz \, (e^{-\nu z} + e^{-\beta_d\nu + \nu z})$$

$$\times \left\{ \left(\frac{\nu}{\mu}\right)^2 z \left(1 - \frac{z}{\beta_d}\right) + \frac{\mu^2 - \nu^2}{\mu^3} \frac{1 + e^{-\beta_d\mu} - e^{-\mu z} - e^{-\beta_d\mu + z\mu}}{1 - e^{-\beta_d\mu}} \right\}. \quad (2.95)$$

The parameters μ and ν can be treated as positive variational parameters that must be chosen to minimize the expression (2.95), thus providing the best possible approximation for the true value of the interaction free energy

$$\frac{F_{\text{int}}^{(P)}}{\hbar\omega}. \quad (2.96)$$

It will turn out that (2.90) and (2.95) result in the inequality

$$\mu^2 - \nu^2 = \frac{K_0^2}{m\omega^2} \geqslant 0. \quad (2.97)$$

Hence the domain of possible values of the parameters μ and ν is given by the inequality

$$\mu \geqslant \nu > 0. \quad (2.98)$$

Let us begin with the simplest (but not the best) choice of variational parameters:

$$\mu = \nu.$$

For this choice, the approximating formula (2.95) is simplified radically and reduces to

$$- \frac{\alpha}{\pi^{1/2}} \int_0^{\beta_d/2} \frac{e^{-z} + e^{-\beta_d + z}}{1 - e^{-\beta_d}} \frac{dz}{\{z(1 - z/\beta_d)\}^{1/2}}. \quad (2.99)$$

It is easy to see that (2.99) represents exactly the first term in the expansion of the free energy $F_{\text{int}}^{(P)}$ in powers of the dimensionless interaction parameter α (i.e. g^2). In fact, thanks to (2.97), the equality $\mu = \nu$ corresponds to the absence of the interaction term in the Hamiltonian $H^{(L)}$, and the functional Φ_0, which is defined by (2.23) and (2.24), is identical zero.

We see also from (2.16) that Φ is proportional to α. Hence the approximation (2.26) contains two terms of expansion for $F_{\text{int}}^{(P)}$ in powers of α, i.e. the zeroth-order term and the first-order term.

The zeroth-order term is now equal to zero and the first-order term is expressed by (2.99). Of course, such an approximation is satisfactory only in the case of small α. But, on the other hand, it is easy to take further steps in this situation and to calculate the term proportional to α^2. This can be done by means of the regular expansion

$$\text{app } F_{\text{int}}^{(P)} = -\vartheta \ln \langle T\{e^{\Phi}\}\rangle_{\Gamma} = A_1 + A_2 + \cdots,$$

$$A_1 = -\vartheta \langle T\{\Phi\}\rangle_{\Gamma}, \tag{2.100}$$

$$A_2 = -\vartheta \frac{1}{2} \langle T\{\Phi^2\} - A_1^2\rangle_{\Gamma}, \ldots$$

Here A_1 is already known, and A_2 can be easily constructed by the method discussed before.

After these remarks, we may turn to (2.95) and consider it from the point of view of the minimum principle, which allows us to determine parameters μ and ν satisfying the inequalities (2.98). To simplify all calculations, we are interested only in the case when

$$\frac{1}{\beta_d} = \frac{\vartheta}{\hbar\omega} \ll 1. \tag{2.101}$$

Substituting $\beta_d = \infty$ into (2.95), we obtain an approximation formula for zero temperature:

$$\text{app } \frac{F_{\text{int}}^{(P)}}{\hbar\omega} = \frac{3(\mu - \nu)}{2} - \frac{3\nu(\mu^2 - \nu^2)}{4}$$

$$\times \int_0^{\infty} e^{-\nu z} \left\{ \left(\frac{\nu}{\mu}\right)^2 z + \frac{\mu^2 - \nu^2}{\mu^3}(1 - e^{-\mu z}) \right\} dz$$

$$- \frac{\alpha}{\pi^{1/2}} \int_0^{\infty} e^{-\nu z} \left\{ \left(\frac{\nu}{\mu}\right)^2 z + \frac{\mu^2 - \nu^2}{\mu^3}(1 - e^{-\mu z}) \right\}^{-1/2} dz. \tag{2.102}$$

After integration, we have

$$\int_0^{\infty} e^{-\nu z} \left\{ \left(\frac{\nu}{\mu}\right)^2 z + \frac{\mu^2 - \nu^2}{\mu^3}(1 - e^{-\mu z}) \right\} dz = \left(\frac{\nu}{\mu}\right)^2 \frac{1}{\nu^2}$$

$$+ \frac{\mu^2 - \nu^2}{\mu^3}\left(\frac{1}{\nu} - \frac{1}{\mu + \nu}\right) = \frac{1}{\mu^2} + \frac{\mu^2 - \nu^2}{\nu\mu^3(\mu + \nu)}\mu = \frac{1}{\mu^2} + \frac{\mu - \nu}{\nu\mu^2} = \frac{1}{\mu\nu}.$$

Thus the two first terms on the right-hand side of (2.102) give us

$$\frac{3}{2}(\mu - \nu) - \frac{3\nu}{4}(\mu^2 - \nu^2)\frac{1}{\mu\nu} = \frac{3}{4\mu}(2\mu^2 - 2\mu\nu - \mu^2 + \nu^2) = \frac{3}{4\mu}(\mu - \nu)^2.$$

Therefore (2.102) results in

$$\text{app } \frac{F_{int}^{(P)}}{\hbar\omega} = \frac{3}{4\mu}(\mu - \nu)^2$$

$$- \frac{\alpha}{\pi^{1/2}} \int_0^\infty e^{-z} \left\{ \left(\frac{\nu}{\mu}\right)^2 z + \frac{\mu^2 - \nu^2}{\mu^3}(1 - e^{-\mu z}) \right\}^{-1/2} dz. \quad (2.103)$$

This formula was derived for the first time by R. Feynman.

Let us investigate the applicability of this formula for the case of small α and try to find parameters ν and μ from the minimum principle in such a way as to take into account terms proportional to α^2. As we have seen (see (2.99)), the term of app $F_{int}^{(P)}$, that is proportional to α in the case $\beta = \infty$ under consideration, will be equal to

$$- \frac{\alpha}{\pi^{1/2}} \int_0^\infty dz \frac{e^{-z}}{z^{1/2}} = -2 \frac{\alpha}{\pi^{1/2}} \int_0^\infty e^{-x^2} dx = -\alpha.$$

This situation corresponds to the choice of parameters $\nu = \mu$. Thus we now put

$$\mu = \nu(1 + \xi\alpha) \quad (2.104)$$

and evaluate the right-hand side of (2.103) with a precision up to terms proportional to α^2. We have

$$\int_0^\infty e^{-z} \frac{dz}{\left\{ z + \frac{\mu^2 - \nu^2}{\mu\nu^2}(1 - e^{-\mu z}) \right\}^{1/2}}$$

$$= \int_0^\infty e^{-z} \frac{dz}{z^{1/2}} - \frac{\mu^2 - \nu^2}{2\mu\nu^2} \int_0^\infty \frac{e^{-z}(1 - e^{-\mu z})}{z^{3/2}} dz + \alpha^2 \ldots$$

$$= 2 \int_0^\infty e^{-x^2} dx - \frac{\mu^2 - \nu^2}{\mu\nu^2} \int_0^\infty \frac{e^{-x^2}(1 - e^{-\mu x^2})}{x^2} dx + \alpha^2 \ldots \quad (2.105)$$

Observing that

$$\frac{\partial}{\partial\mu} \int_0^\infty \frac{e^{-x^2}(1 - e^{-\mu x^2})}{x^2} dx = \int_0^\infty e^{-(1+\mu)x^2} dx = \frac{\pi^{1/2}}{2}(1 + \mu)^{-1/2},$$

we get

$$\int_0^\infty \frac{e^{-x^2}(1 - e^{-\mu x^2})}{x^2} dx = \pi^{1/2}\{(1 + \mu)^{1/2} - 1\}.$$

The relation (2.105) leads to the formula

$$\int_0^\infty e^{-z} \frac{dz}{\left\{ z + \frac{\mu^2 - \nu^2}{\mu\nu^2}(1 - e^{-\mu z}) \right\}^{1/2}}$$

$$= \pi^{1/2} - \frac{\mu^2 - \nu^2}{\mu\nu^2} \pi^{1/2} \{ (1 + \mu)^{1/2} - 1 \} + \alpha^2 \ldots$$

Hence, thanks to (2.104), the right-hand side of this equation can be written as

$$\pi^{1/2}(1 - \alpha\xi D) + \alpha^2 \ldots,$$

where

$$D = \frac{2}{\nu} ((1 + \nu)^{1/2} - 1). \tag{2.106}$$

It follows from (2.103) and (2.104) that

$$\text{app } \frac{F_{\text{int}}^{(P)}}{\hbar\omega} = \frac{3\alpha^2}{4(1 + \alpha\xi)} \nu\xi^2 - \alpha(1 + \alpha\xi)(1 - \alpha\xi D) + \alpha^3 \ldots$$

Neglecting terms of order α^3, we derive the following approximate result:

$$\text{app } \frac{F_{\text{int}}^{(P)}}{\hbar\omega} = -\alpha + \alpha^2 \left(\frac{3}{4} \nu\xi^2 - \xi + \xi D \right). \tag{2.107}$$

Here

$$\xi > 0, \qquad \nu > 0$$

are parameters that must be determined through the minimum principle for the expression

$$\text{app } \frac{F_{\text{int}}^{(P)}}{\hbar\omega} = \text{min.}$$

Let us determine the value of ξ first. We have the extreme condition

$$\frac{\partial}{\partial\xi} \left(\frac{3}{4} \nu\xi^2 - \xi + \xi D \right) = \frac{3}{2} \nu\xi - 1 + D,$$

so that

$$\xi = \frac{2}{3\nu} (1 - D). \tag{2.108}$$

Inserting this value into (2.107), we obtain

$$\text{app } \frac{F_{\text{int}}^{(P)}}{\hbar\omega} = -\alpha - \frac{\alpha^2}{3\nu} (1 - D)^2 = -\alpha - \frac{\alpha^2}{3\nu} \left(1 - \frac{2}{\nu} \{ (1 + \nu)^{1/2} - 1 \} \right)^2. \tag{2.109}$$

Now ν must be determined in such a way that it provides the minimum for the right-hand side of (2.109) or, equivalently, to ensure the maximum of the expression

$$+\frac{1}{\nu} \left(1 - \frac{2}{\nu} \{ (1 + \nu)^{1/2} - 1 \} \right)^2.$$

We can see that the required value is

$$\nu = 3. \tag{2.110}$$

Inserting this value into (2.109), we arrive at the approximation

$$\text{app} \frac{F_{\text{int}}^{(P)}}{\hbar\omega} = -\alpha - \frac{\alpha^2}{81} = -\alpha - 1.23 \left(\frac{\alpha}{10}\right)^2. \tag{2.111}$$

It will be noted that the standard expansion in perturbation theory in powers of α, which has been accomplished up to the second order, gives the following expansion instead of (2.111):

$$-\alpha - 1.26 \left(\frac{\alpha}{10}\right)^2. \tag{2.112}$$

Thus we see that the variational approximation with the simplest choice for the function $E(\nu)$, which corresponds to Feynman's two-body interaction model, provides us with a very precise result in the case of small α.

We now have to consider just the opposite case $\alpha \gg 1$. It is well known that the theory of strong coupling yields in the leading term

$$-0.109\alpha^2. \tag{2.113}$$

Thus, to transform (2.103) in the case of strong interaction, we should apply the following scheme: we assume that μ has order α^2, and ν is proportional to α^0. We obtain from (2.103)

$$\text{app} \frac{F_{\text{int}}^{(P)}}{\hbar\omega} = \frac{3}{4}\mu - \frac{3}{2}\nu + \frac{3}{4}\frac{\nu^2}{\mu}$$

$$- \frac{\alpha}{\pi^{1/2}} \left(\frac{\mu^3}{\mu^2 - \nu^2}\right)^{1/2} \int\limits_0^\infty e^{-z} \left(1 - e^{-\mu z} + \frac{\nu^2\mu}{\mu^2 - \nu^2} z\right)^{-1/2} dz. \tag{2.114}$$

Observing that

$$\frac{\nu^2\mu}{\mu^2 - \nu^2} = \mathcal{O}\left(\frac{1}{\mu}\right) = \mathcal{O}\left(\frac{1}{\alpha^2}\right),$$

we can exploit the expansion

$$\int\limits_0^\infty e^{-z} \left(1 - e^{-\mu z} + \frac{\nu^2\mu}{\mu^2 - \nu^2} z\right)^{-1/2} dz = \int\limits_0^\infty e^{-z}(1 - e^{-\mu z})^{-1/2} dz$$

$$- \frac{\nu^2\mu}{2(\mu^2 - \nu^2)} \int\limits_0^\infty e^{-z} z(1 - e^{-\mu z})^{-3/2} dz + \mathcal{O}\left(\frac{1}{\mu^2}\right) \tag{2.115}$$

and evaluate integrals on the right-hand side of (2.115) by means of the asymptotic formula [i]:

$$\int\limits_0^\infty e^{-z}(1-e^{-\mu z})^{-1/2}\,dz = \int\limits_0^\infty e^{-z}\,dz + \int\limits_0^\infty e^{-z}\{(1-e^{-\mu z})^{-1/2}-1\}\,dz$$

$$= 1 + \frac{1}{\mu}\int\limits_0^\infty e^{-\frac{x}{\mu}}\{(1-e^{-x})^{-1/2}-1\}\,dx$$

$$= 1 + \frac{1}{\mu}\int\limits_0^\infty\{(1-e^{-x})^{-1/2}-1\}\,dx + \mathcal{O}\left(\frac{1}{\mu^2}\right) = 1 + \frac{2\ln 2}{\mu} + \mathcal{O}\left(\frac{1}{\mu^2}\right),$$

$$\int\limits_0^\infty e^{-z}z(1-e^{-\mu z})^{-3/2}\,dz$$

$$= \int\limits_0^\infty e^{-z}z\,dz + \int\limits_0^\infty e^{-z}z\{(1-e^{-\mu z})^{-3/2}-1\}\,dz$$

$$= 1 + \frac{1}{\mu^2}\int\limits_0^\infty e^{-\frac{x}{\mu}}x\{(1-e^{-x})^{-3/2}-1\}\,dx = 1 + \mathcal{O}\left(\frac{1}{\mu^2}\right).$$

The expression (2.115) now gives

$$\int\limits_0^\infty e^{-z}\left(1-e^{-\mu z}+\frac{\nu^2\mu}{\mu^2-\nu^2}z\right)^{-1/2}\,dz = 1 + \frac{2\ln 2-\nu^2/2}{\mu} + \mathcal{O}\left(\frac{1}{\mu^2}\right),$$

and therefore it follows from (2.114) that

$$\text{app}\,\frac{F_{\text{int}}^{(P)}}{\hbar\omega} = \frac{3}{4}\mu - \frac{3}{2}\nu + \frac{3}{4}\frac{\nu^2}{\mu} - \frac{\alpha}{\pi^{1/2}}\mu^{1/2}\left(1+\frac{2\ln 2-\nu^2/2}{\mu}\right) + \mathcal{O}\left(\frac{1}{\alpha^2}\right)$$

$$= \frac{3}{4}\mu - \frac{3}{2}\nu - \frac{\alpha}{\pi^{1/2}}\mu^{1/2}\left(1+\frac{2\ln 2-\nu^2/2}{\mu}\right) + \mathcal{O}\left(\frac{1}{\alpha^2}\right).\quad (2.116)$$

To minimize this expression, let us consider the extreme conditions

$$\frac{\partial}{\partial\nu}\mathcal{E} = 0,\qquad \frac{\partial}{\partial\mu}\mathcal{E} = 0,$$

---

[i] The first integral can be expressed through the Γ-function:

$$\int\limits_0^\infty e^{-z}(1-e^{-\mu z})^{-1/2}\,dz = \frac{1}{\mu}\frac{\Gamma(1/\mu)\sqrt{\pi}}{\Gamma(1/\mu+1/2)}.$$

Introducing the variable $e^{-x} = U$, we get

$$\int\limits_0^\infty\{(1-e^{-x})^{-1/2}-1\}\,dx$$

$$= \int\limits_0^1\{(1-U)^{-1/2}-1\}\frac{dU}{U} = \int\limits_0^1\frac{d}{dU}\left\{\ln\frac{1-(1-U)^{1/2}}{1+(1-U)^{1/2}}-\ln U\right\}\,dU$$

$$= \ln\frac{1-(1-U)^{1/2}}{U\{1+(1-U)^{1/2}\}}\bigg|_0^1 = -\ln\frac{1}{4} = 2\ln 2.$$

where

$$\mathcal{E} = \frac{3}{4}\mu - \frac{3}{2}\nu - \frac{\alpha}{\pi^{1/2}}\mu^{1/2} - \frac{\alpha\mu^{-1/2}}{\pi^{1/2}}2\ln 2 + \frac{\alpha\mu^{-1/2}}{2\pi^{1/2}}\nu^2. \qquad (2.117)$$

These equations give us

$$\nu = \frac{3}{2}\frac{\pi^{1/2}}{\alpha}\mu^{1/2}, \qquad \mu^{1/2} = \frac{2\alpha}{3\pi^{1/2}} - \frac{\alpha\mu^{-1}4\ln 2}{3\pi^{1/2}} + \frac{\alpha\mu^{-1}\nu^2}{3\pi^{1/2}}.$$

From here, it follows that

$$\mu^{1/2} = \frac{2\alpha}{3\pi^{1/2}} + \mathcal{O}\left(\frac{1}{\alpha}\right), \qquad \nu = 1 + \mathcal{O}\left(\frac{1}{\alpha^2}\right).$$

So,

$$\mu^{1/2} = \frac{2\alpha}{3\pi^{1/2}} - \frac{3\pi^{1/2}\ln 2}{\alpha} + \frac{3\pi^{1/2}}{4\alpha} + \mathcal{O}\left(\frac{1}{\alpha^3}\right).$$

Therefore, neglecting terms of order $1/\alpha^2$, we obtain

$$\mu = \frac{4\alpha^2}{9\pi} - 4\ln 2 + 1, \qquad \nu = 1, \qquad (2.118)$$

and the asymptotic expression for Feynman's approximation is

$$\mathcal{E} = -\frac{\alpha^2}{3\pi} - 3\ln 2 - \frac{3}{4} = -0.106\alpha^2 - 2.83 \qquad \text{if} \qquad \alpha \gg 1. \qquad (2.119)$$

The optimal determination of the parameters ν and μ that provides the minimum for the corresponding approximate magnitude of the free energy (2.103) in the case of intermediate values of α needs more elaborate numerical calculations. These calculations have been done (see R. Feynman [5], p. 273). Results of the approximations within the framework of Feynman's model seem to be quite satisfactory when one is interested in the case of statistical equilibrium.

But, as was pointed out by J. T. Devreese and a few other authors, Feynman's model leads to difficulties if it is applied to the investigation of kinetic processes. In fact, this model is characterized by only two parameters: ν and μ. It follows from the previous consideration that the frequency $\nu_0 = \nu\omega$ corresponds to the frequency of the phonon field, while $\mu_0 = \mu\omega$ is the true polaron frequency, determined by the interaction between the electron and the phonon field.

Thus we see that by putting $\nu \neq 1$ we distort the true frequency of the phonon field, which is just equal to ω, not $\nu\omega$. And it is clear that this distortion is essential when collisions between phonons and polarons are considered. Therefore the Hamiltonian $H^{(L)}$ cannot be used to describe kinetic processes when $\nu \neq 1$. One of the possibilities to improve the situation for intermediate values of α consists in a coarser approximation, under which one puts $\nu = 1$ and then minimizes (2.103) with respect to the parameter μ only. In doing so, we find that the approximation

$$-\alpha - 0.98\left(\frac{\alpha}{10}\right)^2.$$

holds for small enough values of α instead of (2.111):

$$-\alpha - 1.23 \left(\frac{\alpha}{10}\right)^2.$$

We believe that the results for the single-frequency model might be improved significantly if proper damping is taken into account.

2.2. Equilibrium Momentum Distribution Function in the Polaron Theory

Let us consider the equilibrium momentum distribution function of the S particle (the electron):

$$W(\mathbf{p}_0) = \frac{\mathrm{Tr}_{S,\Sigma} e^{-\beta H(P)} \delta(\mathbf{p} - \mathbf{p}_0)}{\mathrm{Tr}_{S,\Sigma} e^{-\beta H(P)}}, \qquad (2.120)$$

satisfying the usual normalization condition:

$$\int W(\mathbf{p}_0)\, d\mathbf{p}_0 = 1.$$

Using the Fourier representation for the three-dimensional Dirac δ-function $\delta(\mathbf{p} - \mathbf{p}_0)$, we get

$$W(\mathbf{p}(0)) = \frac{1}{(2\pi)^3} \int \widetilde{W}(\boldsymbol{\lambda}) e^{-i\boldsymbol{\lambda}\cdot\mathbf{p}_0}\, d\boldsymbol{\lambda}, \qquad (2.121)$$

where

$$\widetilde{W}(\boldsymbol{\lambda}) = \frac{\mathrm{Tr}_{S,\Sigma} e^{-\beta H(P)} e^{i\boldsymbol{\lambda}\cdot\mathbf{p}}}{\mathrm{Tr}_{S,\Sigma} e^{-\beta H(P)}}.$$

It should be noted that we have dropped the symbol

$$\lim_{\eta \to 0} \lim_{V \to \infty}$$

here because it will always be clear in future at which step of calculations this passage to the limit is taken. Let us introduce the notation

$$\frac{\mathrm{Tr}_{S,\Sigma} e^{-\beta H(P)} e^{i\boldsymbol{\lambda}\cdot\mathbf{p}}}{\mathrm{Tr}_S e^{-\beta\Gamma}\, \mathrm{Tr}_\Sigma e^{-\beta H(\Sigma)}} = W(\boldsymbol{\lambda}). \qquad (2.122)$$

Then

$$\widetilde{W}(\boldsymbol{\lambda}) = \frac{W(\boldsymbol{\lambda})}{W(0)}. \qquad (2.123)$$

Now we apply to (2.122) the same procedure that was used earlier to transform (2.5) into (2.17). Thus we obtain

$$W(\boldsymbol{\lambda}) = \langle T\{e^{\Phi}\} e^{i\boldsymbol{\lambda}\cdot\mathbf{p}} \rangle_\Gamma.$$

Note that the operators under the T-operation are ordered in such a way that the ordering parameter, say s, increases from the right to the left. One has $s = 0$ at the right end, while $s = \beta\hbar$ at the left end. We also note the identity

$$\mathbf{p} = \mathbf{p}(0).$$

From here, we get

$$T\{e^\Phi\}e^{i\boldsymbol{\lambda}\cdot\mathbf{p}} = T\{e^\Phi\}e^{i\boldsymbol{\lambda}\cdot\mathbf{p}(0)} = T\{e^\Phi e^{i\boldsymbol{\lambda}\cdot\mathbf{p}(0)}\}.$$

Hence

$$W(\boldsymbol{\lambda}) = \langle T\{e^\Phi e^{i\boldsymbol{\lambda}\cdot\mathbf{p}(0)}\}\rangle_\Gamma.$$

Because of this, (2.153) gives

$$\widetilde{W}(\boldsymbol{\lambda}) = \frac{\langle T\{e^\Phi e^{i\boldsymbol{\lambda}\cdot\mathbf{p}(0)}\}\rangle_\Gamma}{\langle T\{e^\Phi\}\rangle_\Gamma}. \tag{2.124}$$

To derive an approximation for (2.124) we follow the same scheme that we implemented in Section 2.1 for the analysis of the free energy $F_{\text{int}}^{(P)}$. In other words, we shall treat formally the difference $\Phi - \Phi_0$, appearing in the expansion

$$T\{e^\Phi ...\} = T\{e^{\Phi_0} ...\} + T\{e^{\Phi_0}(\Phi - \Phi_0)...\} + ...$$

as a magnitude of the "first order of smallness". Then, neglecting in (2.124) terms of higher order of smallness, we get the following approximation:

$$\widetilde{W}_{\text{app}}(\boldsymbol{\lambda}) = \frac{\langle T\{e^{\Phi_0} e^{i\boldsymbol{\lambda}\cdot\mathbf{p}(0)}\}\rangle_\Gamma}{\langle T\{e^{\Phi_0}\}\rangle_\Gamma} + \frac{\langle T\{e^{\Phi_0}(\Phi - \Phi_0)e^{i\boldsymbol{\lambda}\cdot\mathbf{p}(0)}\}\rangle_\Gamma}{\langle T\{e^{\Phi_0}\}\rangle_\Gamma}$$
$$- \frac{\langle T\{e^{\Phi_0} e^{i\boldsymbol{\lambda}\cdot\mathbf{p}(0)}\}\rangle_\Gamma}{\langle T\{e^{\Phi_0}\}\rangle_\Gamma^2} \langle T\{e^{\Phi_0}(\Phi - \Phi_0)\}\rangle_\Gamma. \tag{2.125}$$

Using the auxiliary Hamiltonian $H^{(L)}$, we see easily that

$$\frac{\langle T\{e^{\Phi_0} e^{i\boldsymbol{\lambda}\cdot\mathbf{p}(0)}\}\rangle_\Gamma}{\langle T\{e^{\Phi_0}\}\rangle_\Gamma} = \langle e^{i\boldsymbol{\lambda}\cdot\mathbf{p}}\rangle_{H^{(L)}}.$$

Because $H^{(L)}$ is a positive-definite quadratic form made of Bose operators, we can write

$$\langle e^{i\boldsymbol{\lambda}\cdot\mathbf{p}}\rangle_{H^{(L)}} = e^{-\frac{1}{2}\sum_{\alpha\beta}\langle p_\alpha p_\beta\rangle_{H^{(L)}}\lambda_\alpha\lambda_\beta} = e^{-\frac{\lambda^2\langle\mathbf{p}^2\rangle_{H^{(L)}}}{6}}, \qquad (\alpha,\beta = 1,2,3),$$

and, from here,

$$\frac{\langle T\{e^{\Phi_0} e^{i\boldsymbol{\lambda}\cdot\mathbf{p}(0)}\}\rangle_\Gamma}{\langle T\{e^{\Phi_0}\}\rangle_\Gamma} = \exp\left(-\frac{\lambda^2\langle\mathbf{p}^2\rangle_{H(L)}}{6}\right). \tag{2.126}$$

Let us note that it follows from (1.48) that

$$\frac{1}{3}\langle\mathbf{p}^2\rangle_{H(L)} = \int\limits_{-\infty}^{+\infty} J(\omega)\,d\omega$$

$$= \frac{i\hbar m^2}{2\pi}\int\limits_{-\infty}^{+\infty}\frac{\omega}{1-e^{-\beta\omega\hbar}}\,\frac{1}{m\Omega+\Delta_\infty(\Omega)}\bigg|_{\omega-i0}^{\omega+i0}\,d\omega. \tag{2.127}$$

In the particular case of the single-frequency Σ-system, where

$$\Delta_\infty(\Omega) = -\frac{K_0^2\Omega}{\Omega^2-\nu_0^2},$$

in the notations of (2.89), we have

$$\frac{1}{3}\langle\mathbf{p}^2\rangle_{H(L)} = \frac{m^2/\beta}{m+(K_0/\nu_0)^2} + \frac{K_0^2\hbar}{2\mu_0}\left(\frac{1}{1-e^{-\beta\hbar\mu_0}}+\frac{1}{-1+e^{-\beta\hbar\mu_0}}\right). \tag{2.128}$$

Now we are going to calculate the expression

$$\frac{\langle T\{e^{\Phi_0}(\Phi-\Phi_0)e^{i\boldsymbol{\lambda}\cdot\mathbf{p}(0)}\}\rangle_\Gamma}{\langle T\{e^{\Phi_0}\}\rangle_\Gamma}.$$

Taking into account the definitions of Φ and Φ_0 (see (2.16), (2.23), (2.24) and (2.52)), we get, in full analogy with (2.33),

$$I(\boldsymbol{\lambda}) \equiv \frac{\langle T\{e^{\Phi_0}(\Phi-\Phi_0)e^{i\boldsymbol{\lambda}\cdot\mathbf{p}(0)}\}\rangle_\Gamma}{\langle T\{e^{\Phi_0}\}\rangle_\Gamma} = \frac{1}{4\hbar^2 V}\sum_{(f)}L^2(f)\frac{\hbar}{\omega}\left(1-e^{-\beta\omega\hbar}\right)^{-1}$$

$$\times \int\limits_0^{\hbar\beta} ds \int\limits_0^{\hbar\beta} d\sigma\,\left(e^{-\omega|s-\sigma|}+e^{-\beta\omega\hbar+\omega|s-\sigma|}\right)\frac{\langle T\{e^{\Phi_0+i\mathbf{f}\cdot[\mathbf{R}(s)-\mathbf{R}(\sigma)]+i\boldsymbol{\lambda}\cdot\mathbf{p}(0)}\}\rangle_\Gamma}{\langle T\{e^{\Phi_0}\}\rangle_\Gamma}$$

$$+ \frac{1}{4\hbar^2}\int\limits_0^\infty d\omega' \int\limits_0^{\hbar\beta} ds \int\limits_0^{\hbar\beta} d\sigma\,E_V(\omega')\frac{\hbar\omega'}{1-e^{-\beta\hbar\omega'}}\left(e^{-\omega'|s-\sigma|}+e^{-\beta\omega'\hbar+\omega'|s-\sigma|}\right)$$

$$\times\frac{\langle T\{e^{\Phi_0}[\mathbf{R}(s)-\mathbf{R}(\sigma)]^2 e^{i\boldsymbol{\lambda}\cdot\mathbf{p}(0)}\}\rangle_\Gamma}{\langle T\{e^{\Phi_0}\}\rangle_\Gamma}. \tag{2.129}$$

Note that

$$\frac{\langle T\{e^{\Phi_0}[\mathbf{R}(s)-\mathbf{R}(\sigma)]^2 e^{i\boldsymbol{\lambda}\cdot\mathbf{p}(0)}\}\rangle_\Gamma}{\langle T\{e^{\Phi_0}\}\rangle_\Gamma}$$

$$= -\left\{\sum_{\alpha=1}^3\frac{\partial^2}{\partial f_\alpha^2}\frac{\langle T\{e^{\Phi_0+i\mathbf{f}\cdot[\mathbf{R}(s)-\mathbf{R}(\sigma)]+i\boldsymbol{\lambda}\cdot\mathbf{p}(0)}\}\rangle_\Gamma}{\langle T\{e^{\Phi_0}\}\rangle_\Gamma}\right\}_{\mathbf{f}=0}. \tag{2.130}$$

Generalizing the procedure that led to (2.40) and (2.42), we can derive

$$\frac{\langle T\{e^{\Phi_0+i\mathbf{f}\cdot[\mathbf{R}(s)-\mathbf{R}(\sigma)]+i\boldsymbol{\lambda}\cdot\mathbf{p}(0)}\}\rangle_\Gamma}{\langle T\{e^{\Phi_0}\}\rangle_\Gamma} = \langle T\{e^{i\mathbf{f}\cdot[\mathbf{r}(-is)-\mathbf{r}(-i\sigma)]+i\boldsymbol{\lambda}\cdot\mathbf{p}(0)}\}\rangle_{H^{(L)}}$$

$$= \exp\left(\frac{1}{2}\langle T\{e^{i\mathbf{f}\cdot[\mathbf{r}(-is)-\mathbf{r}(-i\sigma)]+i\boldsymbol{\lambda}\cdot\mathbf{p}(0)}\}^2\rangle_{H^{(L)}}\right). \quad (2.131)$$

Furthermore, we have

$$\langle T\{\{\mathbf{f}\cdot[\mathbf{r}(-is)-\mathbf{r}(-i\sigma)]+i\boldsymbol{\lambda}\cdot\mathbf{p}(0)\}^2\}\rangle_{H^{(L)}}$$

$$= \langle T\{\{\mathbf{f}\cdot[\mathbf{r}(-is)-\mathbf{r}(-i\sigma)]\}^2\}\rangle_{H^{(L)}}$$

$$+ 2\langle T\{\{\mathbf{f}\cdot[\mathbf{r}(-is)-\mathbf{r}(-i\sigma)]\}\boldsymbol{\lambda}\cdot\mathbf{p}(0)\}\rangle_{H^{(L)}} + \frac{\lambda^2}{3}\langle\mathbf{p}^2\rangle_{H^{(L)}}. \quad (2.132)$$

Here $0 < s < \beta\hbar$, $0 < \sigma < \beta\hbar$. As was shown in Section 2.1,

$$\langle T\{\{\mathbf{f}\cdot[\mathbf{r}(-is)-\mathbf{r}(-i\sigma)]\}^2\}\rangle_{H^{(L)}} = \frac{f^2}{3}\mathcal{D}(s-\sigma). \quad (2.133)$$

Here the limiting magnitude is implied on the right-hand side. Note the following important property of $\mathcal{D}$-function:

$$\mathcal{D}(s) = \mathcal{D}(-s), \quad \mathcal{D}(\hbar\beta-s) = \mathcal{D}(s). \quad (2.134)$$

It is worth recalling (1.52) for the averaging procedure with $H^{(L)}$:

$$\langle r_\alpha(t)r_\beta(\tau)\rangle_{H^{(L)}} = \delta_{\alpha\beta}\frac{\hbar i}{2\pi}\int_{-\infty}^{+\infty}\frac{e^{-i\omega(t-\tau)}}{(1-e^{-\beta\omega\hbar})}\frac{1}{m\Omega^2-\eta^2+\Omega\Delta(\Omega)}\Big|_{\omega-i0}^{\omega+i0}d\omega,$$

We note that the function

$$\widetilde{F}(\Omega) = -\frac{1}{m\Omega^2-\eta^2+\Omega\Delta(\Omega)}$$

is invariant with respect to the transformation $\Omega \to -\Omega$:

$$\widetilde{F}(\Omega) = \widetilde{F}(-\Omega).$$

Therefore

$$\widetilde{F}(-\omega+i0) - \widetilde{F}(-\omega-i0) = -[\widetilde{F}(\omega+i0) - \widetilde{F}(\omega-i0)],$$

and we can conclude that

$$\langle r_\alpha(t)r_\beta(\tau)\rangle_{H^{(L)}} = \frac{\hbar i}{2\pi}\delta_{\alpha\beta}\int_0^\infty\left(\frac{e^{-i\omega(t-\tau)}}{1-e^{-\beta\omega\hbar}} - \frac{e^{i\omega(t-\tau)}}{1-e^{\beta\omega\hbar}}\right)\widetilde{F}(\Omega)\Big|_{\omega-i0}^{\omega+i0}d\omega,$$

or, in other words,

$$\langle r_\alpha(t) r_\beta(\tau) \rangle_{H^{(L)}}$$

$$= \delta_{\alpha\beta} \frac{\hbar i}{2\pi} \int_0^\infty \frac{e^{-i\omega(t-\tau)} + e^{-\beta\omega\hbar + i\omega(t-\tau)}}{1 - e^{-\beta\omega\hbar}} \frac{1}{m\Omega^2 - \eta^2 + \Omega\Delta(\Omega)} \bigg|_{\omega - i0}^{\omega + i0} d\omega.$$

It follows from this equation that

$$\langle p_\alpha(t) r_\beta(\tau) \rangle_{H^{(L)}} = m \frac{d}{dt} \langle r_\alpha(t) r_\beta(\tau) \rangle_{H^{(L)}}$$

$$= \delta_{\alpha\beta} \frac{\hbar m}{2\pi} \int_0^\infty \frac{e^{-i\omega(t-\tau)} - e^{-\beta\omega\hbar + i\omega(t-\tau)}}{1 - e^{-\beta\omega\hbar}} \frac{\omega}{m\Omega^2 - \eta^2 + \Omega\Delta(\Omega)} \bigg|_{\omega - i0}^{\omega + i0} d\omega,$$

$$\langle r_\alpha(t) p_\beta(\tau) \rangle_{H^{(L)}}$$

$$= -\delta_{\alpha\beta} \frac{\hbar m}{2\pi} \int_0^\infty \frac{e^{-i\omega(t-\tau)} - e^{-\beta\omega\hbar + i\omega(t-\tau)}}{1 - e^{-\beta\omega\hbar}} \frac{\omega}{m\Omega^2 - \eta^2 + \Omega\Delta(\Omega)} \bigg|_{\omega - i0}^{\omega + i0} d\omega.$$

Taking the usual passage to the thermodynamic limit, we obtain

$$\langle p_\alpha(t) r_\beta(\tau) \rangle_{H^{(L)}}$$

$$= \delta_{\alpha\beta} \frac{\hbar m}{2\pi} \int_0^\infty \frac{e^{-i\omega(t-\tau)} - e^{-\beta\omega\hbar + i\omega(t-\tau)}}{1 - e^{-\beta\omega\hbar}} \frac{1}{m\Omega + \Delta_\infty(\Omega)} \bigg|_{\omega - i0}^{\omega + i0} d\omega, \quad (2.135)$$

$$\langle r_\alpha(t) p_\beta(\tau) \rangle_{H^{(L)}}$$

$$= -\delta_{\alpha\beta} \frac{\hbar m}{2\pi} \int_0^\infty \frac{e^{-i\omega(t-\tau)} - e^{-\beta\omega\hbar + i\omega(t-\tau)}}{1 - e^{-\beta\omega\hbar}} \frac{1}{m\Omega + \Delta_\infty(\Omega)} \bigg|_{\omega - i0}^{\omega + i0} d\omega.$$

Consider the functions

$$F(s) = -\frac{i\hbar m}{2\pi} \int_0^\infty \frac{e^{-\omega s} - e^{-\beta\hbar\omega + \omega s}}{1 - e^{-\beta\omega\hbar}} \frac{1}{m\Omega + \Delta_\infty(\Omega)} \bigg|_{\omega - i0}^{\omega + i0} d\omega \quad (2.136)$$

for $0 < s < \beta\hbar$ and

$$F(s) = \frac{i\hbar m}{2\pi} \int_0^\infty \frac{e^{\omega s} - e^{-\beta\hbar\omega - \omega s}}{1 - e^{-\beta\omega\hbar}} \frac{1}{m\Omega + \Delta_\infty(\Omega)} \bigg|_{\omega - i0}^{\omega + i0} d\omega$$

for $-\beta\hbar < s < 0$ which possess the obvious properties

$$F(-s) = -F(s), \quad -\beta\hbar < s < \beta\hbar,$$

$$F(s + \beta\hbar) = F(s), \quad -\beta\hbar < s < 0, \quad (2.137)$$

$$F(s - \beta\hbar) = F(s), \quad 0 < s < \beta\hbar.$$

If we put $t = -iu$ and $\tau = -is$ in the first relation in (2.135), and $t = -is$, $\tau = -iu$ in the second relation, we get

$$\langle p_\alpha(-iu)r_\beta(-is)\rangle_{H^{(L)}} = i\delta_{\alpha\beta}F(u-s) = -i\delta_{\alpha\beta}F(s-u), \qquad (2.138)$$

$$\langle r_\alpha(-is)p_\beta(-iu)\rangle_{H^{(L)}} = -\delta_{\alpha\beta}F(s-u).$$

Hence, in particular,

$$\langle [r_\alpha(-is) - r_\alpha(-i\sigma)]p_\beta(0)\rangle_{H^{(L)}} = -i\delta_{\alpha\beta}[F(s) - F(\sigma)], \qquad (2.139)$$

from which it follows that

$$\langle \{\mathbf{f} \cdot [\mathbf{r}(-is) - \mathbf{r}(-i\sigma)]\}\boldsymbol{\lambda} \cdot \mathbf{p}(0)\rangle_{H^{(L)}} = -i\mathbf{f} \cdot \boldsymbol{\lambda}[F(s) - F(\sigma)].$$

Thanks to (2.132), (2.133) and (2.139), we have

$$\frac{\langle T\{e^{\Phi_0 + i\mathbf{f}\cdot[\mathbf{R}(s)-\mathbf{R}(\sigma)]+i\boldsymbol{\lambda}\cdot\mathbf{p}(0)}\}\rangle_\Gamma}{\langle T\{e^{\Phi_0}\}\rangle_\Gamma}$$

$$= \exp\left(-\frac{f^2}{6}\mathcal{D}(|s-\sigma|) + i\mathbf{f}\cdot\boldsymbol{\lambda}[F(s) - F(\sigma)] - \frac{\lambda^2}{6}\langle\mathbf{p}^2\rangle_{H^{(L)}}\right). \qquad (2.140)$$

Thus, from (2.130), we find

$$\frac{\langle T\{e^{\Phi_0}[\mathbf{R}(s) - \mathbf{R}(\sigma)]^2 e^{i\boldsymbol{\lambda}\cdot\mathbf{p}(0)}\}\rangle_\Gamma}{\langle T\{e^{\Phi_0}\}\rangle_\Gamma}$$

$$= \left\{\mathcal{D}(|s-\sigma|) + \lambda^2[F(s) - F(\sigma)]^2\right\}e^{-\frac{\lambda^2}{6}\langle\mathbf{p}^2\rangle_{H^{(L)}}}. \qquad (2.141)$$

As a consequence, (2.129) transforms into

$$I(\boldsymbol{\lambda}) \equiv \frac{1}{4\hbar^2}\int d\mathbf{f}\, L^2(f)\frac{\hbar}{\omega(1 - e^{-\beta\omega\hbar})}\int_0^{\hbar\beta} ds\int_0^{\hbar\beta} d\sigma \left(e^{-\omega|s-\sigma|} + e^{-\beta\omega\hbar+\omega|s-\sigma|}\right)$$

$$\times \frac{1}{(2\pi)^3}\exp\left(-\frac{f^2}{6}\mathcal{D}(|s-\sigma|) + i\mathbf{f}\cdot\boldsymbol{\lambda}[F(s) - F(\sigma)] - \frac{\lambda^2}{6}\langle\mathbf{p}^2\rangle_{H^{(L)}}\right)$$

$$+ \frac{1}{4\hbar^2}\int_0^\infty d\omega' \int_0^{\hbar\beta} ds\int_0^{\hbar\beta} d\sigma\, E(\omega')\frac{\omega'\hbar}{1 - e^{-\beta\omega\hbar}}\left(e^{-\omega'|s-\sigma|} + e^{-\beta\omega'\hbar+\omega'|s-\sigma|}\right)$$

$$\times \left\{\mathcal{D}(|s-\sigma|) + \lambda^2[F(s) - F(\sigma)]^2\right\}e^{-\frac{\lambda^2}{6}\langle\mathbf{p}^2\rangle_{H^{(L)}}}. \qquad (2.142)$$

But, because of (2.125), (2.126) and (2.129),

$$\widetilde{W}_{\text{app}}(\boldsymbol{\lambda}) = e^{-\frac{\lambda^2}{6}\langle\mathbf{p}^2\rangle_{H^{(L)}}} + I(\boldsymbol{\lambda}) - I(0)e^{-\frac{\lambda^2}{6}\langle\mathbf{p}^2\rangle_{H^{(L)}}}. \qquad (2.143)$$

Thus it is easy to conclude that

$$\widetilde{W}_{\text{app}}(\boldsymbol{\lambda}) = \left(1 + \frac{1}{4(2\pi)^3} \int df \, \frac{L^2(f)}{\hbar\omega(1 - e^{-\beta\omega\hbar})}\right.$$

$$\times \int_0^{\hbar\beta} ds \int_0^{\hbar\beta} d\sigma \, (e^{-\omega|s-\sigma|} + e^{-\beta\omega\hbar+\omega|s-\sigma|}) e^{-\frac{f^2}{6} \mathcal{D}(|s-\sigma|)}$$

$$\times [\exp\{if \cdot \boldsymbol{\lambda}[F(s) - F(\sigma)]\} - 1]$$

$$+ \frac{\lambda^2}{4\hbar^2} \int_0^\infty d\omega' \int_0^{\hbar\beta} ds \int_0^{\hbar\beta} d\sigma \, E(\omega') \frac{\hbar\omega'}{1 - e^{-\beta\omega\hbar}}$$

$$\left. \times (e^{-\omega'|s-\sigma|} + e^{-\beta\omega'\hbar+\omega'|s-\sigma|})[F(s) - F(\sigma)]^2 \right) e^{-\frac{\lambda^2}{6}\langle\mathbf{p}^2\rangle_{H(L)}}. \quad (2.144)$$

Let us now return to (2.144) for the case of the standard Fröhlich model (2.3), for which

$$L^2(f) = \frac{g^2}{f^2}.$$

Here we can perform an integration of the kind

$$\int df \, F(f^2, \mathbf{f} \cdot \boldsymbol{\lambda}).$$

It is convenient to choose the direction $\boldsymbol{\lambda}$ as the z-axis in f-space. Then

$$\int df \, F(f^2, \mathbf{f} \cdot \boldsymbol{\lambda}) = 2\pi \int_0^\infty df \int_0^\pi d\vartheta \, f^2 F(f^2, \lambda f \cos\vartheta) \sin\vartheta$$

$$= 2\pi \int_0^\infty df \int_{-1}^{+1} dt \, F(f^2, f\lambda t) f^2 = 2\pi \int_{-\infty}^{+\infty} df \int_0^1 dt \, f^2 F(f^2, f\lambda t).$$

Therefore (2.144) yields

$$\widetilde{W}_{\text{app}}(\lambda) = \left(1 + \int_0^{\beta\hbar} ds_1 \int_0^{\beta\hbar} ds_2 \, \frac{g^2}{4(2\pi)^2} \int_{-\infty}^{+\infty} df \int_0^1 dt \, \frac{e^{-\omega|s_1-s_2|} + e^{-\beta\hbar\omega+\omega|s_1-s_2|}}{\hbar\omega(1 - e^{-\beta\omega\hbar})}\right.$$

$$\times e^{-\frac{f^2}{6}\mathcal{D}(s_1-s_2)} \left[\exp\{if\lambda t[F(s_1) - F(s_2)]\} - 1\right]$$

$$+ \frac{\lambda^2}{4\hbar^2} \int_0^{\beta\hbar} ds_1 \int_0^{\beta\hbar} ds_2 \int_0^\infty d\omega' \, E(\omega')\hbar\omega' \frac{e^{-\omega'|s_1-s_2|} + e^{-\beta\hbar\omega'+\omega'|s_1-s_2|}}{1 - e^{-\beta\omega'\hbar}}$$

$$\left. \times [F(s_1) - F(s_2)]^2 \right) e^{-\frac{\lambda^2}{6}\langle\mathbf{p}^2\rangle_{H(L)}}$$

On the other hand,

$$
\int\limits_{-\infty}^{+\infty} df \exp\left(-\frac{f^2}{6}\mathcal{D}(s_1 - s_2) + if\lambda t[F(s_1) - F(s_2)]\right)
$$

$$
= \left(\frac{6\pi}{\mathcal{D}(s_1 - s_2)}\right)^{1/2} e^{-\frac{3}{2}t^2\lambda^2 \frac{[F(s_1)-F(s_2)]^2}{\mathcal{D}(s_1-s_2)}},
$$

and we get

$$
\widetilde{W}_{\mathrm{app}}(\lambda) = \left[1 + \frac{g^2}{4(2\pi)^2\omega\hbar}\int\limits_0^1 dt\int\limits_0^{\beta\hbar} ds_1\int\limits_0^{\beta\hbar} ds_2 \left(\frac{6\pi}{\mathcal{D}(s_1 - s_2)}\right)^{1/2}\right.
$$

$$
\times\frac{e^{-\omega|s_1-s_2|} + e^{-\beta\hbar\omega+\omega|s_1-s_2|}}{1 - e^{-\beta\omega\hbar}}\left(e^{-\frac{3}{2}t^2\lambda^2 \frac{[F(s_1)-F(s_2)]^2}{\mathcal{D}(s_1-s_2)}} - 1\right)
$$

$$
+\frac{\lambda^2}{4\hbar^2}\int\limits_0^{\beta\hbar} ds_1\int\limits_0^{\beta\hbar} ds_2\int\limits_0^{\infty} d\omega'\, E(\omega')\hbar\omega'\frac{e^{-\omega'|s_1-s_2|} + e^{-\beta\hbar\omega'+\omega'|s_1-s_2|}}{1 - e^{-\beta\omega'\hbar}}
$$

$$
\left.\times[F(s_1) - F(s_2)]^2\right]e^{-\frac{\lambda^2}{6}\langle\mathbf{p}^2\rangle_{H^{(L)}}}. \quad (2.145)
$$

Let us now investigate the case of the single-frequency system $\Sigma^{(L)}$ discussed in Section 2.1, for which (2.89) is true and

$$
E(\omega') = \frac{K_0^2}{2}[\delta(\omega' - \nu_0) + \delta(\omega' + \nu_0)], \quad K_0^2 = m(\mu_0^2 - \nu_0^2). \quad (2.146)
$$

For this particular case, (2.145) transforms into

$$
\widetilde{W}_{\mathrm{app}}(\lambda) = \left[1 + \frac{g^2}{4(2\pi)^2\omega\hbar}\int\limits_0^1 dt\int\limits_0^{\beta\hbar} ds_1\int\limits_0^{\beta\hbar} ds_2 \left(\frac{6\pi}{\mathcal{D}(s_1 - s_2)}\right)^{1/2}\right.
$$

$$
\times\frac{e^{-\omega|s_1-s_2|} + e^{-\beta\hbar\omega+\omega|s_1-s_2|}}{1 - e^{-\beta\omega\hbar}}\left(e^{-\frac{3}{2}t^2\lambda^2 \frac{(F(s_1)-F(s_2))^2}{\mathcal{D}(s_1-s_2)}} - 1\right)
$$

$$
+\frac{\lambda^2 m(\mu_0^2 - \nu_0^2)\nu_0}{8\hbar}\int\limits_0^{\beta\hbar} ds_1\int\limits_0^{\beta\hbar} ds_2\frac{e^{-\nu_0|s_1-s_2|} + e^{-\beta\hbar\nu_0+\nu_0|s_1-s_2|}}{1 - e^{-\beta\hbar\nu_0}}
$$

$$
\left.\times [F(s_1) - F(s_2)]^2\right]e^{-\frac{\lambda^2}{6}\langle\mathbf{p}^2\rangle_{H^{(L)}}}. \quad (2.147)
$$

Recall that (see (1.34) and (2.91))

$$\mathcal{D}(s) = \frac{3\hbar}{m}\left[\left(\frac{\nu_0}{\mu_0}\right)^2 |s|\left(1 - \frac{|s|}{\beta\hbar}\right)\right.$$

$$\left. - \frac{\mu_0^2 - \nu_0^2}{\mu_0^3}\frac{1 + e^{-\hbar\beta\mu_0} - e^{-|s|\mu_0} - e^{-\beta\hbar\mu_0 + \mu_0|s|}}{1 - e^{-\beta\hbar\mu_0}}\right], \quad (2.148)$$

$$-\beta\hbar < s < \beta\hbar,$$

$$\frac{1}{3m}\langle \mathbf{p}^2\rangle_{H^{(L)}} = \frac{\nu_0^2}{\beta\mu_0^2} + \frac{\mu_0^2 - \nu_0^2}{2\mu_0}\hbar\frac{1 + e^{-\beta\hbar\mu_0}}{1 - e^{-\beta\hbar\mu_0}}, \quad (2.149)$$

and that in the case under consideration it follows from (2.136) that

$$F(s) = \hbar\left[\left(\frac{\nu_0}{\mu_0}\right)^2\left(\frac{s}{\beta\hbar} - \frac{1}{2}\right) + \frac{\mu_0^2 - \nu_0^2}{2\mu_0^2}\frac{e^{-\beta\hbar\mu_0 + \mu_0 s} - e^{-\mu_0 s}}{1 - e^{-\beta\hbar\mu_0}}\right], \quad (2.150)$$

$$0 < s < \beta\hbar.$$

Introducing the dimensionless parameters and variables

$$\frac{\mu_0}{\omega} = \mu, \quad \frac{\nu_0}{\omega} = \nu, \quad \beta\hbar\omega = \beta_d, \quad \alpha = \frac{g^2}{4\pi\hbar\omega^2}\left(\frac{m}{2\hbar\omega}\right)^{1/2}, \quad (2.151)$$

$$\sigma_1 = \omega s_1, \quad \sigma_2 = \omega s_2$$

we get in the new notation

$$\mathcal{D}(s) = \frac{3\hbar}{m\omega}\mathcal{D}_d(\sigma), \quad F(s) = \hbar F_d(\sigma),$$

$$\mathcal{D}_d(\sigma) = \left(\frac{\nu}{\mu}\right)^2 |\sigma|\left(1 - \frac{|\sigma|}{\beta_d}\right) + \frac{\mu^2 - \nu^2}{\mu^3}\frac{1 + e^{-\beta_d\mu} - e^{-|\sigma|\mu} - e^{-\beta_d\mu + \mu|\sigma|}}{1 - e^{-\beta_d\mu}},$$

$$-\beta_d < \sigma < \beta_d, \quad (2.152)$$

$$F_d(\sigma) = \left(\frac{\nu}{\mu}\right)^2\left(\frac{\sigma}{\beta_d} - \frac{1}{2}\right) + \frac{\mu^2 - \nu^2}{2\mu^2}\frac{e^{-\beta_d\mu + \mu|\sigma|} + e^{-\mu\sigma}}{1 - e^{-\beta_d\mu}},$$

$$0 < \sigma < \beta_d,$$

$$\langle \mathbf{p}^2\rangle_{H^{(L)}} = 3m\omega\hbar\left[\frac{1}{\beta_d}\left(\frac{\nu}{\mu}\right)^2 + \frac{\mu^2 - \nu^2}{2\mu}\frac{1 + e^{-\beta_d\mu}}{1 - e^{-\beta_d\mu}}\right].$$

In this notation, (2.147) takes the form

$$\widetilde{W}_{\mathrm{app}}(\boldsymbol{\lambda}) = \left[1 + \frac{\alpha}{2\pi^{1/2}}\int_0^1 dt \int_0^{\beta_d} d\sigma_1\int_0^{\beta_d} d\sigma_2\left(\frac{1}{\mathcal{D}_d(\sigma_1 - \sigma_2)}\right)^{1/2}\right.$$

$$\left. \times\frac{e^{-|\sigma_1 - \sigma_2|} + e^{-\beta_d + |\sigma_1 - \sigma_2|}}{1 - e^{-\beta_d}}\left(e^{-\frac{t^2\lambda^2}{2}m\omega\hbar\frac{[F_d(\sigma_1) - F_d(\sigma_2)]^2}{\mathcal{D}(\sigma_1 - \sigma_2)}} - 1\right)\right.$$

$$+ \frac{\lambda^2 (m\omega\hbar)}{8} \nu(\mu^2 - \nu^2) \int\limits_0^{\beta_d} d\sigma_1 \int\limits_0^{\beta_d} d\sigma_2 \frac{e^{-\nu|\sigma_1 - \sigma_2|} + e^{-\beta_d\nu + \nu|\sigma_1 - \sigma_2|}}{1 - e^{-\beta_d\nu}}$$

$$\times [F_d(\sigma_1) - F_d(\sigma_2)]^2 \bigg] e^{-\frac{\lambda^2}{6} \langle \mathbf{P}^2 \rangle_{H^{(L)}}}. \quad (2.153)$$

The simplest choice of parameters ν and μ,

$$\nu = \mu = 1, \quad (2.154)$$

corresponds to the case when we put

$$H_{\text{int}}^{(L)} = 0, \quad H^{(L)}(\Sigma) = H(\Sigma), \quad (2.155)$$

and thus

$$\mathbf{r}(-is) = \mathbf{R}(s), \quad \Phi_0 = 0. \quad (2.156)$$

Under such circumstances, the functional Φ in (2.124) is proportional to g^2, i.e. α. Hence we can make use of the formal expansion of the right-hand side of (2.124) in powers of α. It easy to see that if one neglects all terms of second and higher orders in α then (2.153) holds. Therefore the terms in (2.153) are equal to the respective zeroth- and first-order terms in α of this expansion. For the case (2.154) in question, (2.152) gives

$$\mathcal{D}(\sigma) = |\sigma| \left(1 - \frac{|\sigma|}{\beta_d}\right), \quad -\beta_d < \sigma < \beta_d,$$

$$F_d(\sigma_1) - F_d(\sigma_2) = \frac{1}{\beta_d}(\sigma_1 - \sigma_2), \quad 0 < \sigma_1 < \beta_d, \quad 0 < \sigma_2 < \beta_d, \quad (2.157)$$

$$\langle p^2 \rangle_{H^{(L)}} = \frac{3m\omega\hbar}{\beta_d} = 3m\vartheta,$$

from which it follows that

$$\left(\frac{1}{\mathcal{D}_d(\sigma_1 - \sigma_2)}\right)^{1/2} \frac{e^{-|\sigma_1 - \sigma_2|} + e^{-\beta_d + |\sigma_1 - \sigma_2|}}{1 - e^{-\beta_d}} \left(e^{-\frac{t^2\lambda^2}{2} m\omega\hbar \frac{[F_d(\sigma_1) - F_d(\sigma_2)]^2}{\mathcal{D}(\sigma_1 - \sigma_2)}} - 1\right)$$

$$= \Phi(|\sigma_1 - \sigma_2|),$$

$$0 < \sigma_1 < \beta_d, \quad 0 < \sigma_2 < \beta_d,$$

where

$$\Phi(z) = \left[z \left(1 - \frac{z}{\beta_d} \right) \right]^{-1/2} \frac{e^{-z} + e^{-\beta_d + z}}{1 - e^{-\beta_d}} \left(e^{-\frac{t^2 \lambda^2}{2\beta_d^2} m\omega\hbar \frac{z}{1 - z/\beta_d}} - 1 \right),$$

$$0 < z < \beta_d.$$

As long as $\Phi(|\sigma_1 - \sigma_2|)$ is a symmetric function of σ_1 and σ_2, we have

$$\int_0^{\beta_d} d\sigma_1 \int_0^{\beta_d} d\sigma_2 \, \Phi(|\sigma_1 - \sigma_2|)$$

$$= 2 \int_0^{\beta_d} d\sigma_1 \int_0^{\sigma_1} d\sigma_2 \, \Phi(|\sigma_1 - \sigma_2|) = 2 \int_0^{\beta_d} dz \int_0^{\beta_d - z} \Phi(z) \, d\sigma_2$$

$$= 2 \int_0^{\beta_d} (\beta_d - z)\Phi(z) \, dz = 2 \int_0^{\beta_d/2} \Phi(z) \, dz + 2 \int_{\beta_d/2}^{\beta_d} (\beta_d - z)\Phi(z) \, dz$$

$$= 2 \int_0^{\beta_d/2} (\beta_d - z)\Phi(z) \, dz + 2 \int_0^{\beta_d/2} z\Phi(\beta_d - z) \, dz.$$

Hence (2.153) gives

$$\widetilde{W}_I(\lambda) = e^{-\frac{\lambda^2 m\vartheta}{2}} + \frac{\alpha}{\pi^{1/2}} e^{-\frac{\lambda^2 m\omega\hbar}{2\beta_d}} \left[\int_0^1 dt \int_0^{\beta_d/2} dz \, \beta_d \left(\frac{1 - z/\beta_d}{z} \right)^{1/2} \right.$$

$$\times \frac{e^{-z} + e^{-\beta_d + z}}{1 - e^{-\beta_d}} \left(e^{-\frac{t^2 \lambda^2}{2\beta_d^2} m\omega\hbar \frac{z}{1 - z/\beta_d}} - 1 \right)$$

$$\left. + \int_0^1 dt \int_0^{\beta_d/2} dz \left(\frac{z}{1 - z/\beta_d} \right)^{1/2} \frac{e^{-z} + e^{-\beta_d + z}}{1 - e^{-\beta_d}} \left(e^{-\frac{t^2 \lambda^2}{2} m\omega\hbar \frac{1 - z/\beta_d}{z}} - 1 \right) \right].$$

$$(2.158)$$

If $\beta_d \gg 1$ then the terms $e^{-\beta_d/2}$ and $e^{-\beta_d}$ are negligible, and (2.158) becomes

$$\widetilde{W}_I(\lambda) = e^{\frac{\lambda^2 m\omega\hbar}{2\beta_d}} + \frac{\alpha}{\pi^{1/2}} e^{-\frac{\lambda^2 m\omega\hbar}{2\beta_d}} \int_0^1 dt \int_0^{\beta_d/2} dz \left(\frac{1}{z} - \frac{1}{\beta_d} \right)^{1/2} e^{-z} \beta_d$$

$$\times \left(e^{-\frac{t^2 \lambda^2}{2\beta_d^2} m\omega\hbar \frac{z}{1 - z/\beta_d}} - 1 \right) + \frac{\alpha}{\pi^{1/2}} e^{-\frac{\lambda^2 m\omega\hbar}{2\beta_d}}$$

$$\times \int_0^1 dt \int_0^{\beta_d/2} \left(\frac{z}{1 - z/\beta_d} \right)^{1/2} e^{-z} \left(e^{-\frac{t^2 \lambda^2}{2} m\omega\hbar(\frac{1}{z} - \frac{1}{\beta_d})} - 1 \right) dz. \quad (2.159)$$

Taking into account that

$$\int e^{-A\lambda^2 - i\mathbf{p} \cdot \boldsymbol{\lambda}} \, d\boldsymbol{\lambda} = \left(\frac{\pi}{A} \right)^{3/2} e^{-\frac{p^2}{4A}}, \quad A > 0, \quad (2.160)$$

it is easy to derive a momentum partition function in the first order of approximation:

$$W_I(\mathbf{p}) = \frac{1}{(2\pi)^3} \int \widetilde{W}_I(\boldsymbol{\lambda}) e^{-i\boldsymbol{\lambda}\cdot\mathbf{p}}\, d\boldsymbol{\lambda}.$$

For the sake of simplicity, we restrict ourselves to the case of absolute zero temperature, when $\beta_d \to \infty$. Then (2.159) yields

$$\widetilde{W}_I(\boldsymbol{\lambda}) = 1 + \frac{\alpha}{\pi^{1/2}} \int_0^1 dt \int_0^\infty dz\, z^{1/2} e^{-z} \left(e^{-\frac{t^2\lambda^2 m\omega\hbar}{2z}} - 1 \right)$$

$$= 1 - \frac{\alpha}{2} + \frac{\alpha}{\pi^{1/2}} \int_0^1 dt \int_0^\infty dz\, z^{1/2} e^{-z} e^{-\frac{t^2\lambda^2 m\omega\hbar}{2z}}, \qquad (2.161)$$

and hence

$$W_I(\mathbf{p}) = \left(1 - \frac{\alpha}{2}\right)\delta(\mathbf{p})$$

$$+ \frac{1}{(2\pi)^3}\frac{\alpha}{\pi^{1/2}} \int_0^1 dt \int_0^\infty dz\, z^{1/2} e^{-z} e^{-\frac{zp^2}{2t^2 m\omega\hbar}} \left(\frac{2\pi z}{t^2 m\omega\hbar}\right)^{3/2}$$

$$= \left(1 - \frac{\alpha}{2}\right)\delta(\mathbf{p}) + \frac{\alpha}{\pi^2(2m\omega\hbar)^{3/2}} \int_0^1 \frac{dt}{t^3} \int_0^\infty dz\, z^2 e^{-z\left(1+\frac{p^2}{2t^2 m\omega\hbar}\right)}$$

$$= \left(1 - \frac{\alpha}{2}\right)\delta(\mathbf{p}) + \frac{2\alpha}{\pi^2(2m\omega\hbar)^{3/2}} \int_0^1 \frac{dt}{t^3} \frac{1}{\left(1 + p^2/(2t^2 m\omega\hbar)\right)^3}$$

$$= \left(1 - \frac{\alpha}{2}\right)\delta(\mathbf{p}) + \frac{\alpha}{\pi^2(2m\omega\hbar)^{3/2}} \int_1^\infty d\tau \frac{1}{\left(1 + p^2\tau/(2m\omega\hbar)\right)^3}. \qquad (2.162)$$

Thus we arrive at the expression

$$W_I(\mathbf{p}) = \left(1 - \frac{\alpha}{2}\right)\delta(\mathbf{p}) + \frac{\alpha}{(2\pi)^2(2m\omega\hbar)^{1/2}} \frac{1}{p^2\left(1 + p^2/(2m\omega\hbar)\right)^2}. \qquad (2.163)$$

We now see that the approximating Hamiltonian $H^{(L)}$ considered above does not ensure a correct approximation for the partition function $W(\mathbf{p})$, whatever the choice of the spectral function $E(\omega)$. In fact, we always have the equality

$$\langle e^{i\boldsymbol{\lambda}\cdot\mathbf{p}} \rangle_{H^{(L)}} = e^{-\frac{\lambda^2}{6}\langle p^2\rangle_{H^{(L)}}}.$$

Therefore the corresponding momentum partition function will always be of a "nearly Maxwellian type":

$$W_L(\mathbf{p}) = \frac{1}{(2\pi)^3} \int e^{-\frac{\lambda^2}{6}\langle p^2\rangle_{H^{(L)}} - i\boldsymbol{\lambda}\cdot\mathbf{p}}\, d\boldsymbol{\lambda} = \frac{1}{(2\pi)^3}\left(\frac{6\pi}{\langle p^2\rangle_{H^{(L)}}}\right)^{3/2} e^{-\frac{3p^2}{2\langle p^2\rangle_{H^{(L)}}}}.$$

$$(2.164)$$

The choice of $E(\omega)$ will affect only the magnitude of $\langle p^2 \rangle_{H(L)}$ (see (2.127)). For example, if $E(\omega)$ is chosen in the form (2.146) then it follows from (2.148) that

$$\frac{\langle p^2 \rangle_{H(L)}}{m\omega\hbar} = \frac{3}{2}\frac{\mu^2 - \nu^2}{\mu}.$$ (2.165)

And in the first-order approximation (3.153), $W_L(\mathbf{p})$ does not include even the terms of first order in α. Let us return to (2.153). As shown in the "Note" below, we have

$$\delta F(\Gamma) + \delta(\text{app} F_{\text{int}}) = \int W_{\text{int}}(\mathbf{p})\Psi(\mathbf{p})\, d\mathbf{p}\, \delta\xi,$$

i. e.

$$\langle \Psi(\mathbf{p}) \rangle_{\Gamma}\delta\xi + \delta(\text{app} F_{\text{int}}) = \int W_{\text{int}}(\mathbf{p})\Psi(\mathbf{p})\, d\mathbf{p}\, \delta\xi,$$ (2.166)

where the variation is implied to be taken in the form

$$\frac{p^2}{2m} \to \frac{p^2}{2m} + \Psi(\mathbf{p})\delta\xi.$$

Consider the variation of mass

$$\frac{p^2}{2m} \to \frac{p^2}{2(m + \delta m)} = \frac{p^2}{2m} - \frac{p^2}{2m}\delta m.$$

Hence

$$-\left\langle \frac{p^2}{2m^2} \right\rangle_{\Gamma} + \frac{\partial(\text{app} F_{\text{int}})}{\partial m} = -\int W_{\text{app}}(\mathbf{p})\frac{p^2}{2m^2}\, d\mathbf{p}.$$ (2.167)

In particular, for the zero-temperature case ($\vartheta = 0$), we have

$$\left\langle \frac{p^2}{2m^2} \right\rangle_{\Gamma} = 0,$$

so that

$$\int p^2 W_{\text{app}}(\mathbf{p})\, d\mathbf{p} = -2m^2\frac{\partial(\text{app} F_{\text{int}})}{\partial m}.$$

But, as was shown in Section 2.1, if $\vartheta = 0$ then the function

$$\frac{\text{app} F_{\text{int}}}{\hbar\omega}$$

is only a function of α, and, on the other hand, α is proportional to $m^{1/2}$. Whence

$$m\frac{\partial}{\partial m} = \frac{\alpha}{2}\frac{\partial}{\partial \alpha}$$

and

$$\frac{1}{m\omega\hbar}\int p^2 W_{\text{app}}(\mathbf{p})\, d\mathbf{p} = -\alpha\frac{\partial}{\partial \alpha}\left(\frac{\text{app} F_{\text{int}}}{\hbar\omega}\right) \quad \text{for} \quad \vartheta = 0.$$ (2.168)

We also have

$$\frac{1}{m\omega\hbar}\int p^2 W_{\text{app}}(\mathbf{p})\,d\mathbf{p} = -\alpha\frac{\partial}{\partial\alpha}\left(\frac{\text{app}F_{\text{int}}}{\omega\hbar}\right) \quad \text{for} \quad \vartheta = 0. \tag{2.169}$$

Because $(\text{app}F_{\text{int}})$ is a good approximation for the free energy F_{int}, we may conclude that

$$\int p^2 W_{\text{app}}(\mathbf{p})\,d\mathbf{p}$$

will ensure a close-enough approximation for the true expression

$$\int p^2 W(\mathbf{p})\,d\mathbf{p}.$$

However, the quality of the approximation provided by the function $W_{\text{app}}(\mathbf{p})$ for the function $W(\mathbf{p})$ is less satisfactory in some cases. It can be shown, for example, that for small α and $\vartheta = 0$ the function $W_{\text{app}}(\mathbf{p})$ even becomes negative for some region of values of the parameter $p^2/(m\omega\hbar)$ (of the order α).

We should note that (2.143) for $\widetilde{W}(\lambda)$ can be derived immediately from the expression for the free energy. Consider the Hamiltonian

$$H^{(P)} + \delta\Gamma = H^{(P)} + \Psi(\mathbf{p})\,\delta\xi, \tag{2.170}$$

where dx – is an infinitesimal parameter. The corresponding free energy is

$$F(H^{(P)} + \delta\Gamma) = -\vartheta\ln\text{Tr}_{S,\Sigma}e^{-\beta(H^{(P)}+\delta\Gamma)}.$$

Hence

$$\delta F = F(H^{(P)} + \delta\Gamma) - F(H^{(P)}) = \frac{\text{Tr}_{S,\Sigma}e^{-\beta H^{(P)}}\delta\Gamma}{\text{Tr}_{S,\Sigma}e^{-\beta H^{(P)}}} = \langle\delta\Gamma\rangle_{H^{(P)}} =$$

$$= \langle\Psi(\mathbf{p})\rangle_{H^{(P)}}\delta\xi = \int\langle\delta(\mathbf{p}-\mathbf{p}_0)\rangle_{H^{(P)}}\Psi(\mathbf{p}_0)\,d\mathbf{p}_0\,\delta\xi.$$

It follows from (2.120) that

$$\delta F = \int W(\mathbf{p}_0)\Psi(\mathbf{p}_0)\,d\mathbf{p}_0\,\delta\xi. \tag{2.171}$$

Let us say a few words now about the free energy corresponding to the Hamiltonian H' that can be constructed from $H^{(P)}$ by substitution the kinetic energy of the S particle $\mathbf{p}^2/(2m)$ — with a more general function of the momentum $\Gamma'(p)$.

Keeping in mind the method clarified in Section 2.1, which led us to (2.17), we see that the latter does not depend on the particular appearance of the kinetic energy S, so we can write

$$F_{\text{int}}(H') = -\vartheta\ln\langle T\{e^{\Phi}\}\rangle_{\Gamma'}, \tag{2.172}$$

$$F(H') = F_{\text{int}}(H') + F(\Gamma') + F(H(\Sigma)).$$

Here Φ has the same appearance as in Section 2.1 with the only difference that $\mathbf{R}(s)$, which is given by (2.11), has been changed for

$$\mathbf{R}(s) = e^{-\frac{s\Gamma'}{\hbar}} \mathbf{r} e^{-\frac{s\Gamma'}{\hbar}}. \qquad (2.173)$$

It is also useful to note that, in full analogy with (2.38)

$$\frac{\left\langle T\left\{ \exp\left(\Phi_0 + i\int\limits_0^{\hbar\beta} ds\, \tilde{\lambda}(s)\, \mathbf{f}\cdot\mathbf{R}(s) \right) \right\} \right\rangle_{\Gamma'}}{\langle T\{e^{\Phi_0}\}\rangle_{\Gamma'}} = \frac{\mathrm{Tr}_{S,\Sigma} U'(\hbar\beta)}{\mathrm{Tr}_{S,\Sigma} e^{-\beta H'^{(L)}}}. \qquad (2.174)$$

Here $H'^{(L)}$ is the Hamiltonian $H^{(L)}$, in which $p^2/2m$ is changed for the Hamiltonian $\Gamma'(p)$. Φ_0 has the same appearance as in Section 2.1 with $\mathbf{R}(s)$ given by (2.173) and $U(s)$ is determined by the equation generalizing (2.37):

$$\hbar\frac{dU'(s)}{ds} = -[H'^{(L)} - i\hbar\tilde{\lambda}(s)\,\mathbf{f}\cdot\mathbf{r}]U'(s), \qquad (2.175)$$

$$U'(0) = 1.$$

After all these preliminary remarks, we return to the expression (2.172) for the free energy and introduce the approximate expression

$$\mathrm{app}F_{\mathrm{int}}(H') = -\vartheta \ln \langle T\{e^{\Phi_0}\}\rangle_{\Gamma'} - \vartheta \frac{\langle T\{e^{\Phi_0}(\Phi - \Phi_0)\}\rangle_{\Gamma'}}{\langle T\{e^{\Phi_0}\}\rangle_{\Gamma'}}, \qquad (2.176)$$

$$\mathrm{app}F(H') = \mathrm{app}F_{\mathrm{int}}(H') + F(\Gamma') + F(H(\Sigma)). \qquad (2.177)$$

Consider the first-order variation

$$\delta\,\mathrm{app}F = \mathrm{app}F(H^{(P)} + \delta\Gamma) - \mathrm{app}F(H^{(P)})$$

$$= \mathrm{app}F(H^{(P)} + \Psi(\mathbf{p})\delta\xi) - \mathrm{app}F(H^{(P)}). \qquad (2.178)$$

As long as $\delta\xi$ is infinitesimal, this variation will be proportional to $\delta\xi$. The corresponding coefficient is obviously a linear functional $\Psi(\mathbf{p})$. Thus we can write

$$\delta(\mathrm{app}F) = \int f(\mathbf{p}(0))\Psi(\mathbf{p}(0))\,d\mathbf{p}(0)\,\delta\xi.$$

Beginning with (2.171), we shall consider $f(\mathbf{p}(0))$ as an approximation for $W(\mathbf{p}(0))$:

$$\delta(\mathrm{app}F) = \int W_{\mathrm{app}}(\mathbf{p}(0))\Psi(\mathbf{p}(0))\,d\mathbf{p}(0)\,\delta\xi. \qquad (2.179)$$

We intend to show that the Fourier transform

$$\widetilde{W}(\boldsymbol{\lambda}) = \int W_{\text{app}}(\mathbf{p}(0)) e^{i\boldsymbol{\lambda} \cdot \mathbf{p}(0)} \, d\mathbf{p}(0) \tag{2.180}$$

is the same as (2.143) derived before.

Let us start to evaluate (2.178). It must be pointed out here that $H^{(L)}$, upon which $\text{app}F(H^{(L)})$ depends, contains some parameters arising in expressions for $H^{(L)}_{\text{int}}$ and $H^{(L)}(\Sigma)$. These parameters have to be determined from the minimum principle

$$\text{app}F_{\text{int}}(H^P) = \min. \tag{2.181}$$

For example, in Chapter 2 we analyzed the case where the left-hand side of (2.181) depended on two parameters: ν and μ.

In the general case, some other parameters might be included in $\text{app}F_{\text{int}}(H^P)$ through $H^{(L)}_{\text{int}}$ and $H^{(L)}(\Sigma)$. Let us denote them as C_j. With this choice, the left-hand side of (2.181) will be a function of C_j:

$$\text{app}F_{\text{int}}(H^P) = f(...C_j...),$$

and, thanks to the minimum condition, these parameters must satisfy the set of equations

$$\frac{\partial f(...C_j...)}{\partial C_j} = 0.$$

In addition, it follows from (2.177) that the difference

$$\text{app}F(H^{(P)}) - \text{app}F_{\text{int}}(H^{(P)}) = F(\Gamma) + F(H(\Sigma))$$

does not depend on the parameters $(...C_j...)$. Therefore the first-order variation

$$\delta(\text{app}F(H^{(P)})) = \sum_{(j)} \frac{\partial f(...C_j...)}{\partial C_j} \, \delta C_j,$$

which is calculated with respect to the variations of the parameters C_j, is zero. This feature allows us to assume all $...C_j...$ to be fixed when calculating the first-order variation of (2.178).

Let us consider a form

$$F(H^{'(L)}) = -\vartheta \ln \langle T\{e^{\Phi_0}\} \rangle_{\Gamma'} + F(\Gamma') + F(H^{(L)}(\Sigma))$$

and note that, thanks to the variational property mentioned above,

$$F(H^{(L)} + \delta\Gamma) - F(H^{(L)}) = [-\vartheta \ln \langle T\{e^{\Phi_0}\} \rangle_{\Gamma + \delta\Gamma} + F(\Gamma + \delta\Gamma)]$$
$$- [-\vartheta \ln \langle T\{e^{\Phi_0}\} \rangle_{\Gamma} + F(\Gamma)],$$

or, in a brief notation,

$$\delta F(H^{'(L)}) = -\vartheta \delta \ln \langle T\{e^{\Phi_0}\} \rangle_{\Gamma'} + \delta F(\Gamma'). \tag{2.182}$$

On the other hand, the arguments that were used in the derivation of (2.171) lead to

$$\delta F(H^{'(L)}) = \int W_L(\mathbf{p}_0)\Psi(\mathbf{p}_0)\,d\mathbf{p}_0\,\delta\xi, \qquad (2.183)$$

where $W_L(\mathbf{p}_0)$ is the momentum partition function for the particle S of the dynamical system characterized by the Hamiltonian $H^{(L)}$.
Thanks to (2.176), (2.177) and (2.179), we have

$$\int W_{\text{app}}(\mathbf{p}_0)\Psi(\mathbf{p}_0)\,d\mathbf{p}_0\,\delta\xi = \vartheta\delta\ln\langle T\{e^{\Phi_0}\}\rangle_{\Gamma'} + \delta F(\Gamma')$$

$$-\vartheta\delta\frac{\langle T\{e^{\Phi_0}(\Phi-\Phi_0)\}\rangle_{\Gamma'}}{\langle T\{e^{\Phi_0}\}\rangle_{\Gamma'}} = \delta F(H^{'(L)}) - \vartheta\delta\frac{\langle T\{e^{\Phi_0}(\Phi-\Phi_0)\}\rangle_{\Gamma'}}{\langle T\{e^{\Phi_0}\}\rangle_{\Gamma'}}.$$

Bearing (2.183) in mind, we get

$$\delta(F_{\text{app}}) = \int W_{\text{app}}(\mathbf{p}_0)\Psi(\mathbf{p}_0)\,d\mathbf{p}_0\,\delta\xi = \int W_L(\mathbf{p}_0)\Psi(\mathbf{p}_0)\,d\mathbf{p}_0\delta\xi$$

$$-\vartheta\delta\frac{\langle T\{e^{\Phi_0}(\Phi-\Phi_0)\}\rangle_{\Gamma'}}{\langle T\{e^{\Phi_0}\}\rangle_{\Gamma'}}. \qquad (2.184)$$

The definitions of Φ and Φ_0 (see (2.16), (2.23) and (2.24)) make it clear that

$$\delta\frac{\langle T\{e^{\Phi_0}(\Phi-\Phi_0)\}\rangle_{\Gamma'}}{\langle T\{e^{\Phi_0}\}\rangle_{\Gamma'}} = \frac{1}{4\hbar^2 V}\sum_{(f)}L^2(f)\frac{\hbar}{\omega(1-e^{-\beta\omega\hbar})}$$

$$\times\int_0^{\beta\hbar}ds_1\int_0^{\beta\hbar}ds_2\,(e^{-\omega|s_1-s_2|}+e^{-\beta\omega\hbar+\omega|s_1-s_2|})\delta\frac{\langle T\{e^{\Phi_0+i\mathbf{f}\cdot[\mathbf{R}(s_1)-\mathbf{R}(s_2)]}\}\rangle_{\Gamma'}}{\langle T\{e^{\Phi_0}\}\rangle_{\Gamma'}}$$

$$+\frac{1}{4\hbar^2}\int_0^{\infty}d\omega'\,E_V(\omega')\int_0^{\beta\hbar}ds_1\int_0^{\beta\hbar}ds_2\,\frac{\hbar\omega'}{1-e^{-\beta\omega'\hbar}}$$

$$\times\left(e^{-\omega'|s_1-s_2|}+e^{-\beta\omega'\hbar+\omega'|s_1-s_2|}\right)$$

$$\times\left\{-\sum_{\alpha=1}^{3}\frac{\partial^2}{\partial f_\alpha^2}\delta\frac{\langle T\{e^{\Phi_0+i\mathbf{f}\cdot[\mathbf{R}(s_1)-\mathbf{R}(s_2)]}\}\rangle_{\Gamma'}}{\langle T\{e^{\Phi_0}\}\rangle_{\Gamma'}}\right\}_{\mathbf{f}=0}. \qquad (2.185)$$

Introducing a function

$$\widetilde{\lambda}(s) = \delta(s-s_1) - \delta(s-s_2),$$

it follows from (2.174) that

$$\delta \frac{\langle T\{e^{\Phi_0 + i \mathbf{f} \cdot [\mathbf{R}(s_1) - \mathbf{R}(s_2)]}\}\rangle_{\Gamma'}}{\langle T\{e^{\Phi_0}\}\rangle_{\Gamma'}} = \delta \frac{\left\langle T\left\{\exp\left(\Phi_0 + i \int_0^{\hbar\beta} ds\, \widetilde{\lambda}(s)\, \mathbf{f} \cdot \mathbf{R}(s)\right)\right\}\right\rangle_{\Gamma'}}{\langle T\{e^{\Phi_0}\}\rangle_{\Gamma'}}$$

$$= \delta \frac{\mathrm{Tr}_{S,\Sigma}\, U'(\hbar\beta)}{\mathrm{Tr}_{S,\Sigma}\, e^{-\beta H'^{(L)}}}. \quad (2.186)$$

Here

$$\hbar \frac{dU'(s)}{ds} = -[H^{(L)} - i\hbar\widetilde{\lambda}(s)\, \mathbf{f} \cdot \mathbf{r} + \Psi(\mathbf{p})\delta\xi]U'(s),$$

$$U'(0) = 1,$$

and hence

$$U'(s) = U(s) + \delta U(s), \quad (2.187)$$

where

$$\hbar \frac{dU(s)}{ds} = -H^{(L)}U(s) + i\hbar\widetilde{\lambda}(s)\, \mathbf{f} \cdot \mathbf{r}\, U(s), \quad (2.188)$$

$$U(0) = 1, \quad (2.189)$$

and

$$\hbar \frac{d\delta U(s)}{ds} = [-H^{(L)} + i\hbar\widetilde{\lambda}(s)\, \mathbf{f} \cdot \mathbf{r}]\,\delta U(s) - \Psi(\mathbf{p})U(s)\delta\xi, \quad (2.190)$$

$$\delta U(0) = 0.$$

It can also be seen that, thanks to (2.183),

$$\delta \frac{1}{\mathrm{Tr}_{S,\Sigma}\, e^{-\beta H'^{(L)}}} = \delta e^{F(H'^{(L)})} = \beta e^{\beta H'^{(L)}} \delta F(H'^{(L)})$$

$$= \beta e^{\beta F(H^{(L)})} \int W_L(\mathbf{p}_0)\Psi(\mathbf{p}_0)\, d\mathbf{p}_0\, \delta\xi = \frac{\beta \int W_L(\mathbf{p}_0)\Psi(\mathbf{p}(0))\, d\mathbf{p}_0\, \delta\xi}{\mathrm{Tr}_{S,\Sigma}\, e^{-\beta H^{(L)}}}.$$

With the help of (2.186), we obtain

$$\delta \frac{\langle T\{e^{\Phi_0 + i \mathbf{f} \cdot [\mathbf{R}(s_1) - \mathbf{R}(s_2)]}\}\rangle_{\Gamma'}}{\langle T\{e^{\Phi_0}\}\rangle_{\Gamma'}}$$

$$= \frac{\mathrm{Tr}_{S,\Sigma}\, \delta U(\hbar\beta)}{\mathrm{Tr}_{S,\Sigma}\, e^{-\beta H^{(L)}}} + [\mathrm{Tr}_{S,\Sigma}\, U(\hbar\beta)]\,\delta \frac{1}{\mathrm{Tr}_{S,\Sigma}\, e^{-\beta H'^{(L)}}}$$

$$= \frac{\mathrm{Tr}_{S,\Sigma}\, \delta U(\hbar\beta)}{\mathrm{Tr}_{S,\Sigma}\, e^{-\beta H^{(L)}}} + \beta \frac{\mathrm{Tr}_{S,\Sigma}\, U(\hbar\beta)}{\mathrm{Tr}_{S,\Sigma}\, e^{-\beta H^{(L)}}} \int W_L(\mathbf{p}_0)\Psi(\mathbf{p}_0)\, d\mathbf{p}_0\, \delta\xi. \quad (2.191)$$

It follows from (2.188) and (2.189) that

$$U(s) = e^{-s\frac{H^{(L)}}{\hbar}} T\left\{ \exp\left(i \int\limits_0^s \widetilde{\lambda}(\sigma)\, \mathbf{f} \cdot \mathbf{r}(-i\sigma)\, d\sigma \right) \right\}. \qquad (2.192)$$

To solve (2.190), let us consider a solution $U = U(s, u)$ of (2.188) that is equal to the unit operator if $s = u$:

$$\hbar \frac{\partial U(s, u)}{\partial s} = [-H^{(L)} + i\hbar\widetilde{\lambda}(s)\, \mathbf{f} \cdot \mathbf{r}] U(s, u), \quad U(u, u) = 1. \qquad (2.193)$$

Put here

$$U(s, u) = e^{-s\frac{H^{(L)}}{\hbar}} A(s, u) e^{u\frac{H^{(L)}}{\hbar}}.$$

Then

$$\hbar \frac{\partial A(s, u)}{\partial s} = i\hbar\widetilde{\lambda}(s) e^{s\frac{H^{(L)}}{\hbar}}\, \mathbf{f} \cdot \mathbf{r}\, e^{-s\frac{H^{(L)}}{\hbar}} A(s, u) = i\hbar\widetilde{\lambda}(s)\mathbf{f} \cdot \mathbf{r}(-is) A(s, u),$$

$$A(u, u) = 1.$$

The solution of this equation is

$$A(s, u) = T\left\{ \exp\left(i \int\limits_u^s \widetilde{\lambda}(\sigma)\, \mathbf{f} \cdot \mathbf{r}(-i\sigma)\, d\sigma \right) \right\},$$

and

$$U(s, u) = e^{-s\frac{H^{(L)}}{\hbar}} T \exp\left\{ \left(i \int\limits_u^s \widetilde{\lambda}(\sigma)\, \mathbf{f} \cdot \mathbf{r}(-i\sigma)\, d\sigma \right) \right\} e^{u\frac{H^{(L)}}{\hbar}}. \qquad (2.194)$$

Now, it is easy to prove that

$$\delta U(s) = -\frac{1}{\hbar} \int\limits_0^s U(s, u)\Psi(\mathbf{p}) U(u)\, du\, \delta\xi. \qquad (2.195)$$

Really, equation (2.195) gives

$$\hbar \frac{\partial \delta U(s)}{\partial s} = -U(s, s)\Psi(\mathbf{p})U(s)\, \delta\xi - \int\limits_0^s \frac{\partial U(s, u)}{\partial s}\, \Psi(\mathbf{p})U(u)\, du\, \delta\xi,$$

or, thanks to (2.193),

$$\hbar \frac{\partial \delta U(s)}{\partial s}$$

$$= -\Psi(\mathbf{p})U(s)\, \delta\xi - \left(-H^{(L)} + i\hbar\lambda(s)\, \mathbf{f} \cdot \mathbf{r} \right) \frac{1}{\hbar} \int\limits_0^s U(s, u)\Psi(\mathbf{p})U(u)\, du\, \delta\xi.$$

Thus we see that (2.195) satisfies (2.190) with the initial condition

$$\delta U(0) = 0.$$

It follows from (2.195) that

$$\delta U(\beta\hbar) = -\frac{1}{\hbar} \int\limits_0^{\beta\hbar} U(\beta\hbar, u)\Psi(\mathbf{p})U(u)\, du\, \delta\xi. \qquad (2.196)$$

But, thanks to (2.194) and (2.192),

$$U(\beta\hbar, u)\Psi(\mathbf{p})U(u) = e^{-\beta H^{(L)}} T\left\{ \exp\left(i \int_u^{\beta\hbar} \widetilde{\lambda}(\sigma)\, \mathbf{f} \cdot \mathbf{r}(-i\sigma) \right) \right\}$$

$$\times e^{u\frac{H^{(L)}}{\hbar}} \Psi(\mathbf{p}) e^{-u\frac{H^{(L)}}{\hbar}} T\left\{ \exp\left(i \int_0^u d\sigma \widetilde{\lambda}(\sigma)\, \mathbf{f} \cdot \mathbf{r}(-i\sigma) \right) \right\}$$

$$= e^{-\beta H^{(L)}} T\left\{ \exp\left(i \int_u^{\beta\hbar} \widetilde{\lambda}(\sigma)\, \mathbf{f} \cdot \mathbf{r}(-i\sigma) \right) \right\} \Psi(\mathbf{p}(-iu))$$

$$\times T\left\{ \exp\left(i \int_0^u d\sigma\, \widetilde{\lambda}(\sigma)\, \mathbf{f} \cdot \mathbf{r}(-i\sigma) \right) \right\}.$$

Paying attention to the structure of the ordering in the T-products, we can write down this equation in condensed form:

$$U(\beta\hbar, u)\Psi(\mathbf{p})U(u) = e^{-\beta H^{(L)}} T\left\{ \exp\left(i \int_0^{\hbar\beta} d\sigma\, \widetilde{\lambda}(\sigma)\, \mathbf{f} \cdot \mathbf{r}(-i\sigma) \right) \Psi(\mathbf{p}(-iu)) \right\}.$$

$$(2.197)$$

We substitute (2.192), (2.196) and (2.197) into (2.191), recalling that

$$\widetilde{\lambda}(\sigma) = \delta(\sigma - s_1) - \delta(\sigma - s_2).$$

This results in

$$\delta \frac{\langle T\{e^{\Phi_0 + i\mathbf{f}\cdot[\mathbf{R}(s_1) - \mathbf{R}(s_2)]}\}\rangle_{\Gamma'}}{\langle T\{e^{\Phi_0}\}\rangle_{\Gamma'}}$$

$$= -\frac{1}{\hbar} \int_0^{\beta\hbar} du\, \langle T\{e^{\mathbf{f}\cdot[\mathbf{r}(-is_1) - \mathbf{r}(-is_2)]} \Psi(\mathbf{p}(-iu))\}\rangle_{H^{(L)}} \delta\xi$$

$$+ \beta \langle T\{e^{\mathbf{f}\cdot[\mathbf{r}(-is_1) - \mathbf{r}(-is_2)]}\}\rangle_{H^{(L)}} \int W_L(\mathbf{p}_0)\Psi(\mathbf{p}_0)\, d\mathbf{p}_0\, \delta\xi. \quad (2.198)$$

Up to now we have not chosen any explicit form for the function $\Psi(\mathbf{p})$. A possible choice is

$$\Psi(\mathbf{p}) = e^{i\boldsymbol{\lambda}\cdot\mathbf{p}}.$$

Thanks to the definition of $W_L(\mathbf{p})$, it is obvious that

$$\int W_L(\mathbf{p}_0)e^{i\boldsymbol{\lambda}\cdot\mathbf{p}_0}\, d\mathbf{p}_0 = \langle e^{i\boldsymbol{\lambda}\cdot\mathbf{P}} \rangle_{H^{(L)}},$$

and thus

$$\int W_L(\mathbf{p}_0)e^{i\boldsymbol{\lambda}\cdot\mathbf{p}_0}\, d\mathbf{p}_0 = e^{-\frac{\lambda^2}{6}\langle \mathbf{P}^2 \rangle_{H^{(L)}}}. \quad (2.199)$$

This result allows further transformations in (2.198):

$$\langle T\{e^{\mathbf{f}\cdot[\mathbf{r}(-is_1)-\mathbf{r}(-is_2)]}\Psi(\mathbf{p}(-iu))\}\rangle_{H^{(L)}}$$

$$= \langle T\{e^{\mathbf{f}\cdot[\mathbf{r}(-is_1)-\mathbf{r}(-is_2)]+i\boldsymbol{\lambda}\cdot\mathbf{p}(-iu)}\}\rangle_{H^{(L)}}$$

$$= \exp\left(-\frac{1}{2}\langle T\{\{\mathbf{f}\cdot[\mathbf{r}(-is_1)-\mathbf{r}(-is_2)]+\boldsymbol{\lambda}\cdot\mathbf{p}(-iu)\}^2\}\rangle_{H^{(L)}}\right). \quad (2.200)$$

We can also write, as usual (compare with (2.132))

$$\langle T\{\{\mathbf{f}\cdot[\mathbf{r}(-is_1)-\mathbf{r}(-is_2)]+\boldsymbol{\lambda}\cdot\mathbf{p}(-iu)\}^2\}\rangle_{H^{(L)}}$$

$$= \langle T\{\{\mathbf{f}\cdot[\mathbf{r}(-is_1)-\mathbf{r}(-is_2)]\}^2\}\rangle_{H^{(L)}}$$

$$+ 2\langle T\{\{\mathbf{f}\cdot[\mathbf{r}(-is_1)-\mathbf{r}(-is_2)]\}\boldsymbol{\lambda}\cdot\mathbf{p}(-iu)\}\rangle_{H^{(L)}} + \frac{\lambda^2}{3}\langle\mathbf{p}^2\rangle_{H^{(L)}}.$$

But it follows from (2.138) that

$$\langle T\{r_\alpha(-is)p_\beta(-iu)\}\rangle_{H^{(L)}} = -i\delta_{\alpha\beta}F(s-u),$$

with the obvious consequence

$$\langle T\{\{\mathbf{f}\cdot[\mathbf{r}(-is_1)-\mathbf{r}(-is_2)]\}\boldsymbol{\lambda}\cdot\mathbf{p}(-iu)\}\rangle_{H^{(L)}}$$

$$= -i\mathbf{f}\cdot\boldsymbol{\lambda}[F(s_1-u)-F(s_2-u)].$$

Bearing in mind our previous result (2.133), we can reduce (2.200) to the form

$$\langle T\{e^{\mathbf{f}\cdot[\mathbf{r}(-is_1)-\mathbf{r}(-is_2)]}\Psi(\mathbf{p}(-iu))\}\rangle_{H^{(L)}}$$

$$= \exp\left(-\frac{f^2}{6}\mathcal{D}(s_1-s_2)+i\mathbf{f}\cdot\boldsymbol{\lambda}[F(s_1-u)-F(s_2-u)]-\frac{\lambda^2}{6}\langle\mathbf{p}^2\rangle_{H^{(L)}}\right).$$

$$(2.201)$$

We now divide both sides of (2.2) by dx and incorporate (2.198), (2.199) and (2.201) for further transformations. This allows us to derive the following relation:

$$\widetilde{W}_{\text{app}}(\boldsymbol{\lambda}) = e^{-\lambda^2\frac{\langle\mathbf{p}^2\rangle_{H^{(L)}}}{6}} + J(\boldsymbol{\lambda}) - J(0)e^{-\frac{\lambda^2}{6}\langle\mathbf{p}^2\rangle_{H^{(L)}}}, \quad (2.202)$$

where

$$J(\boldsymbol{\lambda}) = \frac{1}{4\hbar^2} \int \frac{df\, L^2(f)}{(2\pi)^3} \frac{1}{\beta\hbar} \int\limits_0^{\beta\hbar} du \int\limits_0^{\beta\hbar} ds_1 \int\limits_0^{\beta\hbar} ds_2 \frac{\hbar(e^{-\omega|s_1-s_2|} + e^{-\beta\omega\hbar+\omega|s_1-s_2|})}{\omega(1-e^{-\beta\omega\hbar})}$$

$$\times \exp\left(-\frac{f^2}{6}\mathcal{D}(s_1 - s_2) + i\mathbf{f}\cdot\boldsymbol{\lambda}\left[F(s_1 - u) - F(s_2 - u)\right] - \frac{\lambda^2}{6}\langle\mathbf{p}^2\rangle_{H(L)}\right)$$

$$+ \frac{1}{4\hbar^2} \int\limits_0^{+\infty} d\omega' \frac{1}{\beta\hbar} \int\limits_0^{\beta\hbar} du \int\limits_0^{\beta\hbar} ds_1 \int\limits_0^{\beta\hbar} ds_2 \frac{\hbar\omega' E(\omega')}{1 - e^{-\hbar\beta\omega'}}$$

$$\times \left(e^{-\omega'|s_1-s_2|} + e^{-\beta\omega'\hbar+\omega'|s_1-s_2|}\right)$$

$$\times \exp\left\{\mathcal{D}(s_1 - s_2) + \lambda^2[F(s_1 - u) - F(s_2 - u)]^2\right\} e^{-\frac{\lambda^2}{6}\langle\mathbf{p}^2\rangle_{H(L)}}. \quad (2.203)$$

It follows from (2.143) that we need only to show that

$$J(\boldsymbol{\lambda}) = I(\boldsymbol{\lambda}) \qquad (2.204)$$

(where $I(\boldsymbol{\lambda})$ is defined by (2.142)) in order to prove the equivalence of (2.203) and (2.143). To compare these expressions, we may represent them in more convenient form:

$$J(\boldsymbol{\lambda}) = \frac{1}{\beta\hbar} \int\limits_0^{\beta\hbar} du \int\limits_0^{\beta\hbar} ds_1 \int\limits_0^{\beta\hbar} ds_2\, \Phi(s_1 - u, s_2 - u|\boldsymbol{\lambda}), \qquad (2.205)$$

$$I(\boldsymbol{\lambda}) = \int\limits_0^{\beta\hbar} ds_1 \int\limits_0^{\beta\hbar} ds_2 \Phi(s_1, s_2|\boldsymbol{\lambda}). \qquad (2.206)$$

But, as has been shown in Section 2.1, the functions

$$\mathcal{D}(s), \quad e^{-\omega|s|} + e^{-\omega\beta\hbar+\omega|s|}, \quad e^{-\omega'|s|} + e^{-\omega'\beta\hbar+\omega|s|}$$

appearing in $\Phi(s_1, s_2)$ which are defined only on the interval $(-\beta\hbar,\ \beta\hbar)$, can be continued to the whole real axis in such a way that they will be periodic functions of s with period $\beta\hbar$.

Equation (2.136) shows that [j]

$$F(s') = F(s'') \quad \text{when } |s' - s''| = \beta\hbar, \quad -\beta\hbar < s', \quad s'' < \beta\hbar.$$

---

[j] Note that, for $0 < s_j < \beta\hbar$ $(j = 1, 2)$, $0 < u < \beta\hbar$, we have $-\beta\hbar < s_j - u < \beta\hbar$.

Hence $F(s)$ can be continued from the corresponding interval $(-\beta\hbar, \beta\hbar)$ to the whole axis $(-\infty, +\infty)$. In addition, this continued function will be a periodic function of s with period $\beta\hbar$. We can put, for example,

$$F(s) = F(s - \beta\hbar) \quad \text{for} \quad \beta\hbar < s < 2\beta\hbar,$$

$$F(s) = F(s + \beta\hbar) \quad \text{for} \quad -2\beta\hbar < s < -\beta\hbar,$$

and so on.

We therefore see that $\Phi(s_1, s_2|\boldsymbol{\lambda})$ can be considered as a periodic function of two arguments s_1 and s_2 with period $\beta\hbar$. In this situation, it is convenient to apply the Fourier transform:

$$\Phi(s_1, s_2|\boldsymbol{\lambda}) = \sum_{(n_1, n_2)} A_{n_1, n_2}(\boldsymbol{\lambda}) e^{i(n_1 s_1 + n_2 s_2)\frac{2\pi}{\beta\hbar}},$$

$$\Phi(s_1 - u, s_2 - u|\boldsymbol{\lambda}) = \sum_{(n_1, n_2)} A_{n_1, n_2}(\boldsymbol{\lambda}) e^{i\frac{2\pi}{\beta\hbar}(n_1 s_1 + n_2 s_2)} e^{i\frac{2\pi}{\beta\hbar}(n_1 + n_2)u},$$

which enables us to derive the relations

$$\int\limits_0^{\beta\hbar} ds_1 \int\limits_0^{\beta\hbar} ds_2\, \Phi(s_1 - u, s_2 - u|\boldsymbol{\lambda}) = (\beta\hbar)^2 A_{0,0}(\boldsymbol{\lambda}) = \int\limits_0^{\beta\hbar} ds_1 \int\limits_0^{\beta\hbar} ds_2\, \Phi(s_1, s_2|\boldsymbol{\lambda}),$$

and

$$\frac{1}{\beta\hbar} \int\limits_0^{\beta\hbar} du \int\limits_0^{\beta\hbar} ds_1 \int\limits_0^{\beta\hbar} ds_2\, \Phi(s_1 - u, s_2 - u|\boldsymbol{\lambda}) = \int\limits_0^{\beta\hbar} ds_1 \int\limits_0^{\beta\hbar} ds_2 \Phi(s_1, s_2|\boldsymbol{\lambda}).$$

Hence (2.205) and (2.206) are equivalent, so that (2.204) is correct. This completes the proof of the equivalence of (2.202) and (2.143).

Chapter 3

KINETIC EQUATIONS IN POLARON THEORY

Ideas and methods put forward by N.N. Bogolubov in [49–51] are outlined and employed in the course of the studies of nonequilibrium polaron properties throughout this chapter. His approach was characterized essentially by an attempt to derive kinetic equations of a physical system rigorously, i. e. without recourse to any approximations based on phenomenological ideas, from the corresponding reversible dynamic equations of classical or quantum mecnanics taken as the starting point. In the work by N.M. Krylov and N.N. Bogolubov [49] a problem of the origin of stochastic behavior in a dynamic system being in weak contact with "large" system was considered.

For classical systems this problem was treated on the basis of the Liouville equation for the probability distribution functions in phase space while the von Neumann equation for statistical density operators was employed in the case of quantum systems [51].

In [49] a method was introduced which allowed derivation of the Fokker-Planck-type equation in the first order approximation. In the monograph [50], published in 1945, methods to derive kinetic equations for "large" systems on the basis of general principles of statistical mechanics were found. In lectures given by N.N. Bogolubov in 1974 at the workshop on statistical mechanics a modified version of the approach [49] was presented an its relation to the theory of two-time correlation Green's functions was discussed [52].

It is worth noticing that the terms "small" and "large" regarding to physical systems are to be understood in that the number of degrees of freedom of the former system is much smaller than this number of the latter one.

Development of ideas outlined in [49–52] enabled formulation of a method of derivation of hierarchical system of formally exact equations for time-dependent averages [35].

An elimination of Bose variables from operator dynamic equations by averaging them out with a properly chosen initial statistical density operator was laid into the foundation of this method. Special lemmas, proved for the case of adiabatic switching on of the interaction between "small" and "large" systems [35], are another cornerstone of this method. In the case when the "large" system is in the state of thermodynamic equilibrium, thus constituting a heat bath, this method allows to describe the process of relaxation to thermodynamic equilibrium for the probability distribution function or statistical density operator of the "small" system. This method also proved itself useful for investigation of superradiant generation processes studied in nonlinear optics [53–55]. Let us also note that this chapter generalizes work by N.N. Bogolubov [33] and outlines approaches to treatment of the electron–phonon system and the elimination of phonon operators from the corresponding kinetic equations.

In particular, a polaron kinetic equation is derived for the interaction of an electron with a phonon field. Moreover, under the proper approximation, the exact Boltzmann equation for the polaron system follows from this kinetic equation. Methods of calculation of the response functions (impedance and admittance), based on the "approximating" Hamiltonian with linear interaction, are also proposed. The equilibrium density probability function of the particle is also calculated.

3.1. Generalized Kinetic Equation.
Method of Rigorous Bose-Amplitude Elimination

Consider a dynamical system S interacting with a phonon field Σ. Let X_S be a set of wave function arguments for a single isolated system S and let $X_\Sigma = (...n_k...)$ stand for the set of the phonon field occupation numbers. Then the dynamical states of the combined system (S, Σ) can be characterized by wave functions of the kind

$$\Psi = \Psi(X_S, X_\Sigma). \tag{3.1}$$

Let us denote by

$$F(t, S), \ f(S) \tag{3.2}$$

operators that, generally speaking, can depend explicitly on time t and act only on the X_S arguments of the wave functions $\Psi(X_S, X_\Sigma)$. Analogously, we denote by

$$G(t, \Sigma), \ g(\Sigma) \tag{3.3}$$

operators that act on the wave function Ψ as a function of the arguments X_Σ. Such operators are, for example, Bose-amplitudes $...b_k...b_k^\dagger....$ It is important to stress that, because $F(t.S)$ and $G(t, S)$ act on different variables of the wave function, they commute with each other. In particular, $F(t, S)$ commutes with all $b_k(t)$ and $b_k^\dagger(t)$. The Hamiltonian of the free phonon field

$$H(\Sigma) = \sum_{(k)} \hbar\omega(k)b_k^\dagger(t)b_k(t), \ \omega(k) > 0. \tag{3.4}$$

represents an example of an operator (3.3). Finally, we denote by

$$\mathcal{U}(t, S, \Sigma)$$

operators acting on either the X_S or X_Σ variables of the wave functions $\Psi(X_S, X_\Sigma)$. Let us remark that all these operators are considered in the Schrödinger representation of dynamical variables. Consider the case when the full Hamiltonian of the system (S, Σ) is, in the notation introduced above,

$$H_t = H(t, S, \Sigma) = \Gamma(t, S) + \sum_{(k)} [C_k(t, S)b_k(t) + C_k^\dagger(t, S)b_k^\dagger(t)] + H(\Sigma),$$

(3.5)

where $\Gamma(t, S)$ — is the free Hamiltonian of the system S and the second term in (3.5) with summation over k describes an interaction between subsystems S and Σ.. Consider two examples of such a system.

I. Polaron theory. The simplest polaron model describes an electron moving through an ionic crystal. The system S consists of one electron placed in an external electric field $\mathcal{E}$:

$$\Gamma(t, S) = \frac{p^2}{2m} + e^{\varepsilon t}\mathbf{E}(t) \cdot \mathbf{r}, \qquad \mathbf{E}(t) = -e\mathcal{E}(t),$$

$$C_k(t, S) = \frac{e^{\varepsilon t}}{V^{1/2}} \mathcal{L}(k) \left(\frac{\hbar}{2\omega(k)}\right)^{1/2} e^{i\mathbf{k}\cdot\mathbf{r}},$$

(3.6)

where e — is the electron charge,

$$\mathcal{L}^*(k) = \mathcal{L}(k),$$

$\mathbf{r}, \mathbf{p}$ — are the position and momentum of the electron, and $\mathcal{L}(k)$, and $\omega(k)$ — are radially symmetric functions of the wave vector $\mathbf{k}$. Summation over k is over the usual quasidiscrete spectrum:

$$k = \left(\frac{2\pi n_1}{L}, \frac{2\pi n_2}{L}, \frac{2\pi n_3}{L}\right), \quad L^3 = V,$$

where n_1, n_2 and n_3 — are integers (positive and negative). Of course, in doing so one keeps in mind the limit $V \to \infty$ leading to the continuous spectrum. The factor $e^{\varepsilon t}$ ($\varepsilon > 0$) is introduced, as usual, to ensure the adiabatic switching on of the interaction. In this case, operators of the type $f(S)$ are functions of operators $\mathbf{p}$, and $\mathbf{r}$, for example

$$f(\mathbf{p}), \quad e^{i\mathbf{k}\cdot\mathbf{r}}, \quad f(\mathbf{p})e^{i\mathbf{k}\cdot\mathbf{r}},$$

and so on. Sometimes we have to use a more general expression for the kinetic energy $T(p)$ instead of $p^2/2m$. Then the Hamiltonian (3.6) must be rewritten as

$$\Gamma(t, S) = T(p) + e^{\varepsilon t}\mathbf{E}(t) \cdot \mathbf{r}. \tag{3.7}$$

II. Fermionic system. The system S is a system of free fermions characterized by the Fermi amplitudes $a_f^\dagger$, and a_f, In this case,

$$\Gamma(t, S) = \sum_{(f)} \Lambda(f) a_f^\dagger a_f, \qquad C_k(t, S) = \frac{e^{\varepsilon t}}{V^{1/2}} L_k \sum_{(f)} a_{f+k}^\dagger a_f,$$

$$C_k^*(t, S) = \frac{e^{\varepsilon t}}{V^{1/2}} L_k^* \sum_{(f)} a_f^\dagger a_{f+k} = \frac{e^{\varepsilon t}}{V^{1/2}} L_k^* \sum_{(f)} a_{f-k}^\dagger a_f, \tag{3.8}$$

where L_k and L_k^* are "c-numbers". Because fermions possess a spin degree of freedom, $f = (\mathbf{f}, \sigma)$,, where the vector $\mathbf{f}$ belongs to the quasidiscrete spectrum and σ is a spin quantum number. The symbol $f + k$ implies that $f + k = (\mathbf{f} + \mathbf{k}, \sigma)$. We also can investigate a system of interacting fermions. In this case, we have to include an interaction operator and terms responsible for the interaction between fermions and external fields in the Hamiltonian $\Gamma(t, S)$.

For dynamical systems of type II, the operators $f(S)$ may be represented as arbitrary combinations of the Fermi amplitudes $...a_f...a_f^\dagger...$, that do not contain any Bose amplitudes, for example $a_{f_1}^\dagger a_{f_2}$. Let us note that problems in the theory of superconductivity and electron transport in metals can be readily reduced to type rimII dynamical systems.

Let us return to the Hamiltonian (3.5) and write the Liouville (von Neumann) equation for the statistical operator D_t of the system (S, Σ):

$$i\hbar \frac{\partial D_t}{\partial t} = H(t, S, \Sigma) D_t - D_t H(t, S, \Sigma) \tag{3.9}$$

with initial condition

$$D_{t_0} = \rho(S) D(\Sigma),$$

$$D(\Sigma) = Z^{-1} e^{-\beta H(\Sigma)}, \quad Z = \text{Tr}_\Sigma \, e^{-\beta H(\Sigma)}, \tag{3.10}$$

$$\text{Tr}_s \, \rho(S) = 1, \quad \text{Tr}_\Sigma D(\Sigma) = 1. \tag{3.11}$$

It can be seen that the initial condition corresponds to the situation where the phonon field Σ is in a state of equilibrium at the time t_0, at which the interaction between the phonon system and the dynamical S system, characterized by the statistical operator $\rho(S)$, is "switched on".

It follows from (3.9) that

$$\text{Tr}_{(S,\Sigma)} D_t = \text{Tr}_{(S,\Sigma)} D_{t_0},$$

and

$$\text{Tr}_{(S,\Sigma)} D_t = \text{Tr}_s \, \rho(S) \, \text{Tr}_\Sigma D(\Sigma) = 1.$$

Thus we have usual normalization condition for the statistical operator D_t of the dynamical system (S, Σ).

Let us introduce an operator $U(t, t_0) = U(t, t_0, S, \Sigma)$, defined by the equation

$$i\hbar \frac{\partial U(t, t_0)}{\partial t} = H(t, S, \Sigma)U(t, t_0), \quad U(t_0, t_0) = 1.$$

Because any Hamiltonian is a Hermitian operator,

$$-i\hbar \frac{\partial U^\dagger(t, t_0)}{\partial t} = U^\dagger(t, t_0)H(t, S, \Sigma), \quad U^\dagger(t_0, t_0) = 1.$$

We can see that U is a unitary operator:

$$U^\dagger(t, t_0) = U^{-1}(t, t_0).$$

With the help of the operators U, we have, from (3.9),

$$D_t = U(t, t_0)D_{t_0}U^{-1}(t, t_0).$$

Consider now some dynamical variable in the Schrödinger representation, $\mathcal{U}(t, S, \Sigma)$. Its average value at t is

$$\langle \mathcal{U} \rangle_t = \mathrm{Tr}_{(S,\Sigma)} \mathcal{U}(t, S, \Sigma)D_t = \mathrm{Tr}_{(S,\Sigma)} \mathcal{U}(t, S, \Sigma)U(t, t_0)D_{t_0}U^{-1}(t, t_0)$$

$$= \mathrm{Tr}_{(S,\Sigma)} \{U^{-1}(t, t_0)\mathcal{U}(t, S, \Sigma)U(t, t_0)\}D_{t_0}. \quad (3.12)$$

It can be seen that the expression

$$U^{-1}(t, t_0)\mathcal{U}(t, S, \Sigma)U(t, t_0) \quad (3.13)$$

is the Heisenberg representation of the dynamical variable $\mathcal{U}(t, S, \Sigma)$, which corresponds to the Schrödinger representation at $t = t_0$. We shall denote this Heisenberg representation by $\mathcal{U}(t, S_t, \Sigma_t)$:

$$\mathcal{U}(t, S_t, \Sigma_t) = U^{-1}(t, t_0)\mathcal{U}(t, S, \Sigma)U(t, t_0). \quad (3.14)$$

In particular, if we consider a dynamical variable given in the Schrödinger representation by the operator $F(t, S)$ then

$$F(t, S_t) = U^{-1}(t, t_0)F(t, S_t)U(t, t_0) = U^\dagger(t, t_0)F(t, S_t)U(t, t_0). \quad (3.15)$$

From (3.12), we get

$$\mathrm{Tr}_{(S,\Sigma)} F(t, S_t)D_{t_0} = \mathrm{Tr}_{(S,\Sigma)} F(t, S)D_t = \mathrm{Tr}_{(S)} F(t, S)(\mathrm{Tr}_{(\Sigma)} D_t).$$

Let us introduce further a statistical operator

$$\rho_t(S) = \mathrm{Tr}_{(\Sigma)} D_t.$$

Then[k]

$$\operatorname{Tr}_{(S,\Sigma)} F(t, S_t) D_{t_0} = \operatorname{Tr}_{(S)} F(t, S)\rho_t(S). \tag{3.16}$$

Consider now a dynamical system described by the Hamiltonian (3.5) with initial condition (3.10) for the statistical operator. Starting from (3.14) for the Heisenberg representation, we have

$$[\mathcal{U}(t, S_t, \Sigma_t), \mathcal{B}(t, S_t, \Sigma_t)] = U^{-1}(t, t_0)[\mathcal{U}(t, S, \Sigma), \mathcal{B}(t, S, \Sigma)]U(t, t_0),$$

where $[\mathcal{U}, \mathcal{B}]$ denotes the commutator

$$[\mathcal{U}, \mathcal{B}] = \mathcal{U}\mathcal{B} - \mathcal{B}\mathcal{U}.$$

From here, we see that if the commutator of two dynamical variables taken in the Schrödinger representation is a c-number then the commutator of these variables in the Heisenberg representation will have the same value.

Denote the Heisenberg representation for the Bose amplitudes as $...b_k(t), ...b_k^\dagger(t)$. Then, according to the definition (3.14),

$$b_k(t_0) = b_k, \quad b_k^\dagger(t_0) = b_k^\dagger.$$

Because the operators $b_k^\dagger$, b_k commute with $\Gamma(t, S)$, $C_k(t, S)$, and $C_k^\dagger(t, S)$, we see that

$$[b_k(t); \Gamma(t, S_t)] = 0, \quad [b_k^\dagger(t); \Gamma(t, S_t)] = 0,$$

$$[b_k(t); C_k(t, S_t)] = 0, \quad [b_k^\dagger(t); C_k(t, S_t)] = 0, \tag{3.17}$$

$$[b_k(t); C_k^\dagger(t, S_t)] = 0, \quad [b_k^\dagger(t); C_k^\dagger(t, S_t)] = 0.$$

For the same reason,

$$[H(\Sigma_t); f(S_t)] = 0. \tag{3.18}$$

---

[k] By definition, the average value at the moment t is
$$\langle F \rangle_t = \operatorname{Tr} F(t, S) \cdot D_t(S, \Sigma).$$
The trace is calculated over the complete set of the states of the system $S + \Sigma$. Taking the tensor products $|S\rangle \otimes |\sigma\rangle$ of the base states $|S\rangle$ of the system S by the base states $|\sigma\rangle$ of the system Σ as the base states $|s, \sigma\rangle$ of the whole system $S + \Sigma$, let us write down $\langle F \rangle_t$ as
$$\langle F \rangle_t = \operatorname{Tr}_S F(t, S) \cdot \operatorname{Tr}_\Sigma D_t(S, \Sigma) = \operatorname{Tr}_{(S)} F(t, S) \cdot \rho_t(S),$$
where $\rho_t = \operatorname{Tr}_\Sigma D_t(S, \Sigma)$ is the density matrix of the S-system.

It is clear that $b_k^\dagger(t)$ and $b_k(t)$ satisfy the same commutation relations as $b_k^\dagger$ and b_k. Bearing in mind (3.5) and (3.17), we can write dynamical equations for the Bose amplitudes:

$$i\hbar\frac{\partial b_k(t)}{\partial t} = [b_k(t), H(t, S_t, \Sigma_t)],$$

$$i\hbar\frac{\partial b_k(t)}{\partial t} = \hbar\omega(k)b_k(t) + C_k^\dagger(t, S_t),$$

i. e.

$$\frac{\partial b_k(t)}{\partial t} = -i\omega(k)b_k(t) - \frac{i}{\hbar}C_k^\dagger(t, S_t),$$

The conjugate equation is

$$\frac{\partial b_k^\dagger(t)}{\partial t} = i\omega(k)b_k^\dagger(t) + \frac{i}{\hbar}C_k(t, S_t).$$

Taking into account the initial conditions, we can write the formal solution of these equations:

$$b_k(t) = \widetilde{b}_k(t) - i\mathcal{B}_k(t),$$

$$\widetilde{b}_k(t) = e^{-i\omega(k)(t-t_0)}b_k, \tag{3.19}$$

$$\mathcal{B}_k(t) = \frac{1}{\hbar}\int_{t_0}^{t} d\tau\, e^{-i\omega(k)(t-\tau)}C_k^\dagger(\tau, S_\tau),$$

and also

$$b_k^\dagger(t) = \widetilde{b}_k^\dagger(t) + i\mathcal{B}_k^\dagger(t),$$

$$\widetilde{b}_k^\dagger(t) = e^{i\omega(k)(t-t_0)}b_k^\dagger, \tag{3.20}$$

$$\mathcal{B}_k^\dagger(t) = \frac{1}{\hbar}\int_{t_0}^{t} d\tau\, e^{i\omega(k)(t-\tau)}C_k(\tau, S_\tau).$$

Let us consider a dynamical variable, which can be expressed in the Schrödinger representation by the explicitly time-independent operator $f(S)$. The equation of motion for $f(S_t)$ follows from (3.5) and (3.17):

$$i\hbar\frac{\partial f(S_t)}{\partial t} = [f(S_t), H(t, S_t, \Sigma_t)]$$

This equation can be rewritten in the explicit form

$$i\hbar\frac{\partial f(S_t)}{\partial t} = [f(S_t), \Gamma(t, S_t)]$$

$$+ \sum_{(k)}b_k(t)[f(S_t), C_k(t, S_t)] + \sum_{(k)}b_k^\dagger(t)[f(S_t), C_k^\dagger(t, S_t)].$$

Substituting (3.19) and (3.20) into the last equation and taking the trace over all variables $\text{Tr}_{(S,\Sigma)}...D_{t_0}$, we get

$$i\hbar\frac{\partial}{\partial t}\text{Tr}_{(S,\Sigma)}f(S_t)D_{t_0} + \text{Tr}_{(S,\Sigma)}[\Gamma(t,S_t), f(S_t)]D_{t_0}$$

$$= -i\sum_{(k)}\text{Tr}_{(S,\Sigma)}\mathcal{B}_k(t)[f(S_t), C_k(t,S_t)]D_{t_0}$$

$$+ i\sum_{(k)}\text{Tr}_{(S,\Sigma)}\mathcal{B}_k^\dagger(t)[f(S_t), C_k^\dagger(t,S_t)]D_{t_0}$$

$$+ \sum_{(k)}\text{Tr}_{(S,\Sigma)}\widetilde{b}_k(t)[f(S_t), C_k(t,S_t)]D_{t_0}$$

$$+ \sum_{(k)}\text{Tr}_{(S,\Sigma)}\widetilde{b}_k^\dagger(t)[f(S_t), C_k^\dagger(t,S_t)]D_{t_0}. \quad (3.21)$$

In order to get rid of the Bose amplitudes $\widetilde{b}_k$ and $\widetilde{b}_k^\dagger$ on the right-hand side of (3.21), we formulate the following lemma.

Lemma For average values of the product of two operators $\widetilde{b}_k(t)$ and $\mathcal{U}(S, \Sigma)$, the following relations hold:

$$\text{Tr}_{(S,\Sigma)}\widetilde{b}_k(t)\mathcal{U}(S,\Sigma)D_{t_0} = (1+N_k)\text{Tr}_{(S,\Sigma)}\{\widetilde{b}_k(t)\mathcal{U}(S,\Sigma) - \mathcal{U}(S,\Sigma)\widetilde{b}_k(t)\}D_{t_0},$$

where

$$N_k = \frac{e^{-\beta\hbar\omega(k)}}{1 - e^{-\beta\hbar\omega(k)}},$$

$$\text{Tr}_{(S,\Sigma)}\mathcal{U}(S,\Sigma)\widetilde{b}_k(t)D_{t_0} = N_k\text{Tr}_{(S,\Sigma)}\{\widetilde{b}_k(t)\mathcal{U}(S,\Sigma) - \mathcal{U}(S,\Sigma)\widetilde{b}_k(t)\}D_{t_0}.$$

For the proof, see Appendix I.

Choosing $\mathcal{U}(S, \Sigma) = [f(S_t), C_k(t, S_t)]$, we derive the useful relations

$$\text{Tr}_{(S,\Sigma)}\widetilde{b}_k(t)[f(S_t), C_k(t,S_t)]D_{t_0}$$

$$= (1+N_k)\text{Tr}_{(S,\Sigma)}[\widetilde{b}_k(t), [f(S_t), C_k(t,S_t)]]D_{t_0},$$

$$\text{Tr}_{(S,\Sigma)}\widetilde{b}_k^\dagger(t)[f(S_t), C_k^\dagger(t,S_t)]D_{t_0}$$

$$= N_k\text{Tr}_{(S,\Sigma)}[[f(S_t), C_k^\dagger(t,S_t)], \widetilde{b}_k^\dagger(t)]D_{t_0}. \quad (3.22)$$

Because the operators $...b_k...b_k^\dagger...$ commute with $[f(S_t)C_k(t,S_t)]$, and $[f(S_t), C_k^\dagger(t,S_t)]$,

$$[b_k(t), [f(S_t), C_k(t,S_t)]] = 0, \qquad [[f(S_t), C_k^\dagger(t,S_t)], b_k^\dagger(t)] = 0.$$

Substituting (3.19) and (3.20) into these identities, we find that

$$[\tilde{b}_k(t), [f(S_t), C_k(t, S_t)]] = i[\mathcal{B}_k(t), [f(S_t), C_k(t, S_t)]]$$

$$= i\mathcal{B}_k(t)[f(S_t), C_k(t, S_t)] - i[f(S_t), C_k(t, S_t)]\mathcal{B}_k(t),$$

$$[[f(S_t), C_k^\dagger(t, S_t)], \tilde{b}_k^\dagger(t)] = -i[[f(S_t), C_k^\dagger(t, S_t)], \mathcal{B}_k^\dagger(t)]$$

$$\text{(3.23)}$$

$$= i\mathcal{B}_k^\dagger(t)[f(S_t), C_k^\dagger(t, S_t)] - i[f(S_t), C_k^\dagger(t, S_t)]\mathcal{B}_k^\dagger(t).$$

Making use of (3.22) and (3.23), we derive from (3.21)

$$i\hbar\frac{\partial}{\partial t}\text{Tr}_{(S,\Sigma)}f(S_t)D_{t_0} + \text{Tr}_{(S,\Sigma)}[\Gamma(t, S_t), f(S_t)]D_{t_0}$$

$$+ \sum_{(k)}i\Big\{N_k\text{Tr}_{(S,\Sigma)}\mathcal{B}_k(t)[f(S_t), C_k(t, S_t)]D_{t_0}$$

$$+ (1 + N_k)\text{Tr}_{(S,\Sigma)}[C_k(t, S_t), f(S_t)]\mathcal{B}_k(t)D_{t_0}\Big\}$$

$$+ i\sum_{(k)}\Big\{(1 + N_k)\text{Tr}_{(S,\Sigma)}\mathcal{B}_k^\dagger(t)[f(S_t), C_k^\dagger(t, S_t)]D_{t_0}$$

$$+ N_k\text{Tr}_{(S,\Sigma)}[C_k^\dagger(t, S_t), f(S_t)]\mathcal{B}_k^\dagger(t)D_{t_0}\Big\}. \quad \text{(3.23a)}$$

Note that, thanks to (3.16),

$$\text{Tr}_{(S,\Sigma)}f(S_t)D_{t_0} = \text{Tr}_{(S)}f(S)\rho_t(S),$$

$$\text{Tr}_{(S,\Sigma)}[\Gamma(t, S_t), f(S_t)]D_{t_0} = \text{Tr}_{(S)}\{\Gamma(t, S)f(S) - f(S)\Gamma(t, S)\}\rho_t(S).$$

Substituting the operators $b_k(t)$ and $b_k^\dagger(t)$ in (3.23a) with their explicit expressions (3.19) and (3.20), and dividing both sides of the resulting equation by $i\hbar$, we find that

$$\text{Tr}_{(S)}\left(f(S)\frac{\partial\rho_t(S)}{\partial t} + \frac{\Gamma(t, S)f(S) - f(S)\Gamma(t, S)}{i\hbar}\rho_t(S)\right)$$

$$= \frac{1}{\hbar^2}\sum_{(k)}\int_{t_0}^t d\tau\,\text{Tr}_{(S,\Sigma)}e^{-i\omega(k)(t-\tau)}\{N_kC_k^\dagger(\tau, S_\tau)[f(S_t), C_k(t, S_t)]$$

$$+ (1 + N_k)[C_k^\dagger(t, S_t), f(S_t)]C_k(\tau, S_\tau)\}D_{t_0}$$

$$+ \frac{1}{\hbar^2}\sum_{(k)}\int_{t_0}^t d\tau\,\text{Tr}_{(S,\Sigma)}e^{i\omega(k)(t-\tau)}\{(1 + N_k)C_k(\tau, S_\tau)[f(S_t), C_k^\dagger(t, S_t)]$$

$$+ N_k[C_k^\dagger(t, S_t), f(S_t)]C_k(\tau, S_\tau)\}D_{t_0}. \quad \text{(3.23')}$$

Thus we have constructed the generalized kinetic equation. Now we pass to the consideration of the polaron model itself, which was proclaimed to be the main goal of this chapter. Substituting (3.6), i.e.

$$\Gamma(t, S) = T(p) + e^{\varepsilon t}\mathbf{E}(t) \cdot \mathbf{r}, \qquad \mathbf{E}(t) = -e\mathbf{E}(t),$$

$$C_k(t, S) = \frac{e^{\varepsilon t}}{V^{1/2}} \mathcal{L}(k) \left(\frac{\hbar}{2\omega(k)}\right)^{1/2} e^{i\mathbf{k}\cdot\mathbf{r}},$$

into the right-hand side of the generalized kinetic equation, we find that

$$\mathrm{Tr}_{(S)} \left\{ f(S)\frac{\partial \rho_t(S)}{\partial t} + \frac{e^{\varepsilon t}\mathbf{E}(t) \cdot [\mathbf{r}f(S) - f(S)\mathbf{r}] + T(p)f(S) - f(S)T(p)}{i\hbar} \rho_t(S) \right\}$$

$$= \frac{1}{V} e^{2\varepsilon t} \sum_{(k)} \frac{\mathcal{L}^2(k)}{2\hbar\omega(k)} \int_{t_0}^{t} d\tau\, e^{-\varepsilon(t-\tau)} [N_k e^{-\omega(k)(t-\tau)} + (1 + N_k)e^{\omega(k)(t-\tau)}]$$

$$\times \mathrm{Tr}_{(S,\Sigma)} \{ e^{-i\mathbf{k}\cdot\mathbf{r}_\tau} f(S_t)e^{i\mathbf{k}\cdot\mathbf{r}_t} - e^{-i\mathbf{k}\cdot\mathbf{r}_\tau} e^{i\mathbf{k}\cdot\mathbf{r}_t} f(S_t) \} D_{t_0}$$

$$+ \frac{1}{V} e^{2\varepsilon t} \sum_{(k)} \frac{\mathcal{L}^2(k)}{2\hbar\omega(k)} \int_{t_0}^{t} d\tau\, e^{-\varepsilon(t-\tau)} [(1 + N_k)e^{-\omega(k)(t-\tau)} + N_k e^{\omega(k)(t-\tau)}]$$

$$\times \mathrm{Tr}_{(S,\Sigma)} \{ e^{i\mathbf{k}\cdot\mathbf{r}_t} f(S_t)e^{-i\mathbf{k}\cdot\mathbf{r}_\tau} - f(S_t)e^{i\mathbf{k}\cdot\mathbf{r}_t} e^{-i\mathbf{k}\cdot\mathbf{r}_\tau} \} D_{t_0}, \qquad (3.24)$$

It is interesting to observe that the operators of the phonon field do not enter this equation explicitly. The right-hand side of the equation depends only on the electron trajectory.

Let us stress that the electron operators $\mathbf{r}(\tau)$ and $\mathbf{p}(\tau)$, $(t_0 \leqslant \tau \leqslant t)$ depend on the initial values $\mathbf{r}, \mathbf{p}, ..., b_k, b_k^\dagger$ in a very complicated manner. Therefore, in order to derive some relevant results from (3.24) we have to restrict ourselves to a proper approximation, assuming, for example, that $f(S) = f(\mathbf{p})$ and substituting the intricate time dependence of the electron trajectory $\mathbf{r}_\tau$ with the uniform-motion trajectory

$$\mathbf{r}(\tau) = \mathbf{r}(t) - \frac{\mathbf{p}(t)}{m}(t - \tau),$$

considered as the "zeroth-order approximation". Within the framework of the Fröhlich model, taking into account the smallness of the electron–phonon interaction parameter, one can derive explicitly the approximate Boltzmann equation for the polaron. This equation contains an integral term induced only by the one-phonon emission and absorption processes.

Consider the spatially uniform case, i.e the case when $f(S) = f(\mathbf{p})$ and hence $f(S_t) = f(\mathbf{p}_t)$. From the usual quantum mechanical commutative rules, we have

$$\mathbf{r}f(\mathbf{p}) - f(\mathbf{p})\mathbf{r} = i\hbar \frac{\partial f(\mathbf{p})}{\partial \mathbf{p}}.$$

It is obvious that

$$\text{Tr}_{(S)} f(\mathbf{p})\rho_t(S) = \int f(\mathbf{p})W_t(\mathbf{p})\,d\mathbf{p},$$

where

$$W_t(\mathbf{p}) = \text{Tr}_{(S)} \delta(\mathbf{p} - \mathbf{p}_0)\rho_t(S). \tag{3.25}$$

Let $\mathbf{p}_t$ be a momentum operator in the Heisenberg representation. Then, with the help of (3.16)

$$\text{Tr}_{(S,\Sigma)} F(\mathbf{p}_t)D_{t_0} = \int F(\mathbf{p})W_t(\mathbf{p})\,d\mathbf{p}.$$

It follows from (3.11) and (3.25) that

$$\int W_t(\mathbf{p})\,d\mathbf{p} = 1.$$

It is clear that $W_t(\mathbf{p})$ may be interpreted as the probability density at time t. The left-hand side of (3.24) can be represented as

$$\text{Tr}_{(S)}\left(f(\mathbf{p})\frac{\partial\rho_t(S)}{\partial t} + e^{\varepsilon t}\mathbf{E}(t)\cdot\frac{\partial f(\mathbf{p})}{\partial \mathbf{p}}\,\rho_t(S)\right)$$

$$= \int d\mathbf{p}\, f(\mathbf{p})\left(\frac{\partial W_t(\mathbf{p})}{\partial t} + e^{\varepsilon t}\mathbf{E}(t)\cdot\frac{\partial f(\mathbf{p})}{\partial \mathbf{p}}\,W_t(\mathbf{p})\right). \tag{3.26}$$

It is easily seen that

$$e^{i\mathbf{k}\cdot\mathbf{r}}f(\mathbf{p}) = f(\mathbf{p} - \hbar\mathbf{k})e^{i\mathbf{k}\cdot\mathbf{r}}, \qquad f(\mathbf{p})e^{i\mathbf{k}\cdot\mathbf{r}} = e^{i\mathbf{k}\cdot\mathbf{r}}f(\mathbf{p} + \hbar\mathbf{k}), \tag{3.27}$$

and also

$$e^{i\mathbf{k}\cdot\mathbf{r}_t}f(\mathbf{p}_t) = f(\mathbf{p}_t - \hbar\mathbf{k})e^{i\mathbf{k}\cdot\mathbf{r}_t}, \qquad f(\mathbf{p}_t)e^{i\mathbf{k}\cdot\mathbf{r}_t} = e^{i\mathbf{k}\mathbf{r}_t}f(\mathbf{p}_t + \hbar\mathbf{k}).$$

Taking into account the invariance of both sides of (3.24) with respect to the transformation $\mathbf{k} \to -\mathbf{k}$ and the notes made above about the probability density function, we find that

$$\int d\mathbf{p}\, f(\mathbf{p})\left(\frac{\partial W_t(\mathbf{p})}{\partial t} - e^{\varepsilon t}\mathbf{E}(t)\cdot\frac{\partial W_t(\mathbf{p})}{\partial \mathbf{p}}\right)$$

$$= \int d\mathbf{p}\, f(\mathbf{p})\left(\frac{\partial W_t(\mathbf{p})}{\partial t} + e^{\varepsilon t}\mathbf{E}(t)\cdot\frac{\partial f(\mathbf{p})}{\partial \mathbf{p}}\,W_t(\mathbf{p})\right)$$

$$= \frac{1}{V}e^{2\varepsilon t}\sum_{(k)}\frac{\mathcal{L}^2(k)}{2\hbar\omega(k)}\int_{t_0}^t d\tau\, e^{-\varepsilon(t-\tau)}[N_k e^{-\omega(k)(t-\tau)} + (1+N_k)e^{\omega(k)(t-\tau)}]$$

$$\times \text{Tr}_{(S,\Sigma)}\left\{e^{i\mathbf{k}\cdot\mathbf{r}_\tau}e^{-i\mathbf{k}\cdot\mathbf{r}_t}[f(\mathbf{p}_t - \hbar\mathbf{k}) - f(\mathbf{p})]D_{t_0}\right\}$$

$$+ \frac{1}{V}e^{2\varepsilon t}\sum_{(k)}\frac{\mathcal{L}^2(k)}{2\hbar\omega(k)}\int_{t_0}^t d\tau\, e^{-\varepsilon(t-\tau)}[(1+N_k)e^{-\omega(k)(t-\tau)} + N_k e^{\omega(k)(t-\tau)}]$$

$$\times \text{Tr}_{(S,\Sigma)}\left\{[f(\mathbf{p}_t - \hbar\mathbf{k}) - f(\mathbf{p})]e^{i\mathbf{k}\cdot\mathbf{r}_t}e^{-i\mathbf{k}\cdot\mathbf{r}_\tau}D_{t_0}\right\}, \tag{3.28}$$

where $D_{t_0} = \rho(S)D(\Sigma)$. The rigorous equation (3.28) will be considered in the next section as a source for the derivation of various approximate kinetic equations.

Let us note in conclusion that the generalized equation (3.23′) can be used in other applications. For example, it can be applied to investigate the motion of electrons in a metal if one derives the corresponding kinetic equations. For this purpose, one must put in (3.23′)

$$f(S) = a_f^\dagger a_f, \qquad \Gamma = \sum_{(f)} T_f a_f^\dagger a_f.$$

Then

$$\mathrm{Tr}_{(S)}\, f(S)\rho_t(S) = \mathrm{Tr}_{(S)}\, a_f^\dagger a_f \rho_t(S) = \langle a_f^\dagger(t)a_f(t)\rangle_{t_0} = n_f(t)$$

and

$$\mathrm{Tr}_{(S)}\, f(S)\frac{\partial \rho_t(S)}{\partial t} = \frac{\partial}{\partial t}\, n_f(t),$$

and the expression

$$C_k(t, S) = \frac{\exp(\varepsilon t)}{V^{1/2}} \mathcal{L}_k \sum_{(f)} a_{f+k}^\dagger a_f(t)$$

would stand for the operator $C_k(t, S)$. Let us define the operators $a_{f+k}^\dagger(t)$ and $a_f(t)$ entering this combination, assuming that they satisfy the following approximate equation of motion without interaction:

$$i\hbar\frac{da_f}{dt} = T_f a_f(t).$$

From here,

$$a_f(\tau) = \exp\left(-i\frac{T_f}{\hbar}(\tau - t)\right) a_f(t), \qquad a_f^\dagger(\tau) = \exp\left(i\frac{T_f}{\hbar}(\tau - t)\right) a_f^\dagger(t).$$

Thus

$$C_k(\tau, S_\tau) = \frac{\exp(\varepsilon\tau)}{V^{1/2}} \sum_{(f)} \exp\left(-i\frac{T_{f+k} - T_f}{k}(t - \tau)\right) a_{f+k}^\dagger(t)a_f(t).$$

Taking into account the discussion above and substituting the approximate expression for $C_k(\tau, S_\tau)$ into the generalized kinetic equation (3.23′), after simple transformations and the standard passage to the limit $t_0 \to -\infty$, $\varepsilon \to 0$, we arrive at the well-known Bloch quantum kinetic equation, the basic equation in the theory of the electrical and thermal conductivity of metals and semiconductors [48].

3.2. Kinetic Equations in the First-Order Approximation for Weak Interactions

In this section we are concerned with the weak-interaction case. It is convenient to characterize the electron-phonon interaction strength by a small dimensionless parameter, denoted by α, under the assumption that $\mathcal{L}^2(k)$ is proportional to α. For example, within the framework of the Fröhlich model,

$$\alpha = \frac{g^2}{4\pi\hbar\omega^2}\left(\frac{m}{2\hbar\omega}\right)^{1/2}. \tag{3.29}$$

is generally adopted as the standard dimensionless parameter. We also assume that the external force $\mathbf{E}$ is formally proportional to some small parameter.

In the "zeroth-order approximation", when we neglect the electron–phonon interaction, the following equation of motion holds:

$$i\hbar\frac{d\mathbf{r}}{d\tau} = \mathbf{r}T(p) - T(p)\mathbf{r}, \tag{3.30}$$

from which it follows that

$$\mathbf{r}_\tau = e^{\frac{i}{\hbar}T(p)(\tau-\tau_0)}\mathbf{r}_{\tau_0}e^{-\frac{i}{\hbar}T(p)(\tau-\tau_0)}.$$

Let $\tau_0 = t$; then

$$\mathbf{r}_t = e^{\frac{i}{\hbar}T(p)(\tau-t)}\mathbf{r}_{\tau_0}e^{-\frac{i}{\hbar}T(p)(\tau-t)},$$

and

$$e^{i\mathbf{k}\cdot\mathbf{r}_\tau} = \exp\left(\frac{i}{\hbar}T(p_t)(\tau-t)\right)e^{i\mathbf{k}\cdot\mathbf{r}_t}\exp\left(-\frac{i}{\hbar}T(p_t)(\tau-t)\right). \tag{3.31}$$

Moving $e^{i\mathbf{k}\cdot\mathbf{r}_t}$ to the right in (3.31) with the help of (3.27), we obtain

$$e^{i\mathbf{k}\cdot\mathbf{r}_\tau} = e^{\frac{i}{\hbar}T(\mathbf{p}_t)(\tau-t)}e^{-\frac{i}{\hbar}T(\mathbf{p}_t-\hbar\mathbf{k})(\tau-t)}e^{i\mathbf{k}\cdot\mathbf{r}_t} = e^{\frac{i}{\hbar}[T(\mathbf{p}_t)-T(\mathbf{p}_t-\hbar\mathbf{k})](\tau-t)}e^{i\mathbf{k}\cdot\mathbf{r}_t}, \tag{3.32a}$$

and also

$$e^{i\mathbf{k}\cdot\mathbf{r}_\tau} = e^{i\mathbf{k}\cdot\mathbf{r}_t}e^{\frac{i}{\hbar}[T(\mathbf{p}_t+\hbar\mathbf{k})-T(\mathbf{p}_t)](\tau-t)}. \tag{3.32b}$$

Under the transformation $\mathbf{k} \to -\mathbf{k}$, we have

$$e^{-i\mathbf{k}\cdot\mathbf{r}_\tau} = e^{-i\mathbf{k}\cdot\mathbf{r}_t}e^{\frac{i}{\hbar}[T(\mathbf{p}_t-\hbar\mathbf{k})-T(\mathbf{p}_t)](\tau-t)}.$$

This "approximation" will be used in (3.28) only for the terms proportional to α.

We substitute (3.32a) and (3.32b) under the trace operation, exploiting the "zeroth-order approximation" in the following manner:

$$\mathcal{E}_{\text{app}} = \left\{\text{Tr}_{(S,\Sigma)}\,e^{i\mathbf{k}\cdot\mathbf{r}_\tau}e^{-i\mathbf{k}\cdot\mathbf{r}_t}[f(\mathbf{p}_t-\hbar\mathbf{k}) - f(\mathbf{p}_t)]D_{t_0}\right\}_{\text{app}}$$

$$= \text{Tr}_{(S,\Sigma)}\,e^{\frac{i}{\hbar}[T(\mathbf{p}_t)-T(\mathbf{p}_t-\hbar\mathbf{k})(\tau-t)}[f(\mathbf{p}_t-\hbar\mathbf{k}) - f(\mathbf{p}_t)]D_{t_0}, \tag{3.33}$$

$$\mathcal{E}_{\text{app}}^* = \left\{ \text{Tr}_{(S,\Sigma)}[f(\mathbf{p}_t - \hbar\mathbf{k}) - f(\mathbf{p}_t)]e^{i\mathbf{k}\cdot\mathbf{r}_t}e^{-i\mathbf{k}\cdot\mathbf{r}_\tau} \right\}_{\text{app}}$$

$$= \text{Tr}_{(S,\Sigma)}[f(\mathbf{p}_t - \hbar\mathbf{k}) - f(\mathbf{p}_t)]e^{\frac{i}{\hbar}[T(\mathbf{p}_t - \hbar\mathbf{k}) - T(\mathbf{p}_t)](\tau - t)}D_{t_0}.$$

It should be pointed out that all of these expressions are multiplied by the magnitude $\mathcal{L}(k)$, which is proportional to α.

Thus we suppose that the terms of first order on the right-hand side of (3.28) are evaluated correctly. This is just the approximation we have been striving for. Further, take the limit $V \to \infty$, $t_0 \to -\infty$, and then put $\varepsilon \to 0$. in the final results. First of all, however, we have to transform the expressions (3.33) for $\mathcal{E}_{\text{app}}^*$ and $\mathcal{E}_{\text{app}}$. Let us return to the relation

$$\text{Tr}_{(S,\Sigma)} F(\mathbf{p}_t) D_{t_0} = \int F(\mathbf{p}) W_t(\mathbf{p}) \, d\mathbf{p},$$

which holds for an arbitrary function of the momentum, $F(\mathbf{p})$. For $F(\mathbf{p})$ we choose

$$F(\mathbf{p}) = e^{\frac{i}{\hbar}[T(\mathbf{p}_t) - T(\mathbf{p}_t - \hbar\mathbf{k})](\tau - t)}[f(\mathbf{p}_t - \hbar\mathbf{k}) - f(\mathbf{p}_t)],$$

and we have, as a result,

$$\mathcal{E}_{\text{app}}e = \int d\mathbf{p} \, e^{\frac{i}{\hbar}[T(\mathbf{p}) - T(\mathbf{p} - \hbar\mathbf{k})](\tau - t)}[f(\mathbf{p} - \hbar\mathbf{k}) - f(\mathbf{p})]W_t(\mathbf{p})$$

$$= \int_{(\mathbf{p} \to \mathbf{p} + \hbar\mathbf{k})} d\mathbf{p} \, e^{-\frac{i}{\hbar}[T(\mathbf{p} + \hbar\mathbf{k}) - T(\mathbf{p})](\tau - t)} f(\mathbf{p})W_t(\mathbf{p} + \hbar\mathbf{k})$$

$$- \int d\mathbf{p} \, e^{\frac{i}{\hbar}[T(\mathbf{p}) - T(\mathbf{p} - \hbar\mathbf{k})](\tau - t)} f(\mathbf{p})W_t(\mathbf{p}).$$

It is easily seen that $\mathcal{E}_{\text{app}}^*$ is the complex conjugate of $\mathcal{E}_{\text{app}}$:

$$\mathcal{E}_{\text{app}}^* = \int d\mathbf{p} \, e^{-\frac{i}{\hbar}[T(\mathbf{p} + \hbar\mathbf{k}) - T(\mathbf{p})](\tau - t)} f(\mathbf{p})W_t(\mathbf{p} + \hbar\mathbf{k})$$

$$- \int d\mathbf{p} \, e^{\frac{i}{\hbar}[T(\mathbf{p} - \hbar\mathbf{k}) - T(\mathbf{p})](\tau - t)} f(\mathbf{p})W_t(\mathbf{p}).$$

We substitute these expressions into (3.28), taking the limit $V \to \infty$, and changing all sums $V^{-1}\sum_{(k)}(...)$ to integrals $(2\pi)^{-3} \int d\mathbf{k}$. It is also convenient to make the transformation $\mathbf{k} \to -\mathbf{k}$ in the integrals containing $W_t(\mathbf{p})$. Let us introduce a new variable of integration $t - \tau = \xi$, so that

$$\int_{t_0}^{t} d\tau \, (...) = \int_{0}^{t - t_0} d\xi \, (...).$$

In the limit $t_0 \to -\infty$, these integrals take the form $\int_0^\infty d\xi \, (...)$. In this way, we can derive the first-order approximation for (3.28):

$$\int d\mathbf{p} \, f(\mathbf{p}) \left(\frac{\partial W_t(\mathbf{p})}{\partial t} - e^{\varepsilon t}\mathbf{E}(t) \cdot \frac{\partial W_t(\mathbf{p})}{\partial \mathbf{p}} \right)$$

$$= \frac{e^{2\varepsilon t}}{(2\pi)^3} \int d\mathbf{p} \, f(\mathbf{p}) \int d\mathbf{k} \, \frac{\mathcal{L}^2(k)}{2\hbar\omega(k)} \, A_\varepsilon(\mathbf{p}, \mathbf{k}),$$

where

$$A_\varepsilon(\mathbf{p}, \mathbf{k}) = \int\limits_0^\infty d\xi\, e^{-\varepsilon\xi}[(1 + N_k)e^{i\omega(k)\xi} + N_k e^{-i\omega(k)\xi}]$$
$$\times [e^{-i\xi\Delta_{p,k}} W_t(\mathbf{p} + \hbar\mathbf{k}) - e^{i\xi\Delta_{p,k}} W_t(\mathbf{p})]$$
$$+ \int\limits_0^\infty d\xi\, e^{-\varepsilon\xi}[(1 + N_k)e^{-i\omega(k)\xi} + N_k e^{i\omega(k)\xi}]$$
$$\times [e^{i\xi\Delta_{p,k}} W_t(\mathbf{p} + \hbar\mathbf{k}) - e^{-i\xi\Delta_{p,k}} W_t(\mathbf{p})]$$

and where

$$\Delta_{p,k} = \frac{T(\mathbf{p} + \hbar\mathbf{k}) - T(\mathbf{p})}{\hbar}.$$

In view of the fact that $f(\mathbf{p})$ is arbitrary function of the momentum, $\mathbf{p}$, this equation can be reduced to an equation for the probability density function $W_t(\mathbf{p})$:

$$\frac{\partial W_t(\mathbf{p})}{\partial t} - e^{\varepsilon t}\mathbf{E}(t) \cdot \frac{\partial W_t(\mathbf{p})}{\partial \mathbf{p}} = \frac{e^{2\varepsilon t}}{(2\pi)^3} \int d\mathbf{k}\, \frac{\mathcal{L}^2(k)}{2\hbar\omega(k)}\, A_\varepsilon(\mathbf{p}, \mathbf{k}). \qquad (3.34)$$

Collecting similar terms in the expression for $A_\varepsilon(\mathbf{p}, \mathbf{k})$, we find that

$$A_\varepsilon(\mathbf{p}, \mathbf{k}) = [(1 + N_k)W_t(\mathbf{p} + \hbar\mathbf{k}) - N_k W_t(\mathbf{p})]$$
$$\times \left(\int\limits_0^\infty e^{-\varepsilon\xi} e^{-\xi[\Delta_{p,k} - \omega(k)]}\, d\xi + \int\limits_0^\infty e^{-\varepsilon\xi} e^{i\xi[\Delta_{p,k} - \omega(k)]}\, d\xi \right)$$
$$+ [N_k W_t(\mathbf{p} + \hbar\mathbf{k}) - (1 + N_k)W_t(\mathbf{p})]$$
$$\times \left(\int\limits_0^\infty e^{-\varepsilon\xi} e^{-\xi[\Delta_{p,k} + \omega(k)]}\, d\xi + \int\limits_0^\infty e^{-\varepsilon\xi} e^{i\xi[\Delta_{p,k} + \omega(k)]}\, d\xi \right).$$

Here

$$N_k = \frac{e^{-\beta\hbar\omega(k)}}{1 - e^{-\beta\hbar\omega(k)}}.$$

or, that is

$$A_\varepsilon(\mathbf{p}, \mathbf{k}) = \frac{W_t(\mathbf{p} + \hbar\mathbf{k}) - e^{-\beta\hbar\omega(k)}W_t(\mathbf{p})}{1 - e^{-\beta\hbar\omega(k)}}\, D_\varepsilon(\Delta_{p,k} - \omega(k))$$
$$+ \frac{W_t(\mathbf{p} + \hbar\mathbf{k})e^{-\beta\hbar\omega(k)} - W_t(\mathbf{p})}{1 - e^{-\beta\hbar\omega(k)}}\, D_\varepsilon(\Delta_{p,k} + \omega(k)),$$
$$D_\varepsilon(z) = \int\limits_{-\infty}^{+\infty} e^{-\varepsilon|\xi|} e^{i\xi z}\, d\xi.$$

Let us also observe that

$$\lim_{\varepsilon \to 0} D_\varepsilon(\Delta_{p,k} \mp \omega(k)) = 2\pi\delta(\Delta_{p,k} \mp \omega(k)) = 2\pi\delta\left(\frac{\hbar\Delta_{p,k} \mp \hbar\omega(k)}{\hbar}\right)$$

$$= 2\pi\hbar\delta(\hbar\Delta_{p,k} \mp \hbar\omega(k)),$$

Therefore

$$\lim_{\varepsilon \to 0} A_\varepsilon(\mathbf{p}, \mathbf{k}) = \frac{2\pi\hbar}{1 - e^{-\beta\hbar\omega(k)}} [W_t(\mathbf{p} + \hbar\mathbf{k}) - e^{-\beta\hbar\omega(k)}W_t(\mathbf{p})]$$

$$\times \delta(T(\mathbf{p} + \hbar\mathbf{k}) - T(\mathbf{p}) - \hbar\omega(k))$$

$$+ \frac{2\pi\hbar}{1 - e^{-\beta\hbar\omega(k)}} [W_t(\mathbf{p} + \hbar\mathbf{k})e^{-\beta\hbar\omega(k)} - W_t(\mathbf{p})]$$

$$\times \delta(T(\mathbf{p} + \hbar\mathbf{k}) - T(\mathbf{p}) + \hbar\omega(k)).$$

Now, let us take the final step, putting $\varepsilon \to 0$ in (3.34). As a result, we get the final form of the kinetic equation in the first-order approximation:

$$\frac{\partial W_t(\mathbf{p})}{\partial t} - \mathbf{E}(t) \cdot \frac{\partial W_t(\mathbf{p})}{\partial \mathbf{p}} = \frac{1}{(2\pi)^2} \int d\mathbf{k} \, \frac{\mathcal{L}^2(k)}{2\omega(k)(1 - e^{-\beta\hbar\omega(k)})}$$

$$\times [W_t(\mathbf{p} + \hbar\mathbf{k}) - e^{-\beta\hbar\omega(k)}W_t(\mathbf{p})]\delta(T(\mathbf{p} + \hbar\mathbf{k}) - T(\mathbf{p}) - \hbar\omega(k))$$

$$+ \frac{1}{(2\pi)^2} \int d\mathbf{k} \, \frac{\mathcal{L}^2(k)}{2\omega(k)(1 - e^{-\beta\hbar\omega(k)})}$$

$$\times [W_t(\mathbf{p} + \hbar\mathbf{k})e^{-\beta\hbar\omega(k)} - W_t(\mathbf{p})]\delta(T(\mathbf{p} + \hbar\mathbf{k}) - T(\mathbf{p}) + \hbar\omega(k)). \quad (3.35)$$

Thus we have obtained generalized Boltzmann equations. Consider now an important particular case,

$$T(p) = \frac{p^2}{2m}.$$

Consequently, all δ-functions take the form

$$\delta\left(\frac{(\mathbf{p} + \hbar\mathbf{k})^2}{2m} - \frac{\mathbf{p}^2}{2m} \pm \hbar\omega(k)\right).$$

It is obvious that (3.35) will be the usual Boltzmann equation, in which the integral terms on the right-hand side correspond to one-phonon emission and absorption. Such a Boltzmann equation has been studied intensely in the investigation of transport properties.

If the electric field is time-independent then the stationary Boltzmann equation reads as

$$-\mathbf{E}(t) \cdot \frac{\partial W(\mathbf{p})}{\partial \mathbf{p}} = \frac{1}{(2\pi)^2} \int d\mathbf{k} \, \frac{\mathcal{L}^2(k)}{2\omega(k)(1 - e^{-\beta\hbar\omega(k)})}$$

$$\times [W(\mathbf{p} + \hbar\mathbf{k}) - e^{-\beta\hbar\omega(k)} W(\mathbf{p})]\delta\left(\frac{(\mathbf{p} + \hbar\mathbf{k})^2}{2m} - \frac{\mathbf{p}^2}{2m} - \hbar\omega(k)\right)$$

$$+ \frac{1}{(2\pi)^2} \int d\mathbf{k} \, \frac{\mathcal{L}^2(k)}{2\omega(k)(1 - e^{-\beta\hbar\omega(k)})}$$

$$\times [W(\mathbf{p} + \hbar\mathbf{k})e^{-\beta\hbar\omega(k)} - W(\mathbf{p})]\delta\left(\frac{(\mathbf{p} + \hbar\mathbf{k})^2}{2m} - \frac{\mathbf{p}^2}{2m} + \hbar\omega(k)\right). \quad (3.36)$$

The factor $e^{-\beta\hbar\omega(k)}$ in (3.36) can be omitted in the case of low temperatures. The resulting equation was analyzed by Devreese and Evrard [6] for the Fröhlich polaron model. Very complicated behavior of the stationary probability density function $W(\mathbf{p})$ was revealed, apparently indicating the existence of a fundamental peculiarity at $E = 0$.

In conclusion we should like to say a few words about one approximation used to determine the relation between the applied electric field and the average stationary electron velocity $\mathcal{V}$.

We multiply both sides of (3.36) by $\mathbf{p}$ and integrate over the whole of momentum space. After simple transformations, we find that

$$-\mathbf{E} = \frac{1}{(2\pi)^2} \int d\mathbf{k} \, \frac{\mathcal{L}^2(k)\hbar\mathbf{k}}{2\omega(k)(1 - e^{-\beta\hbar\omega(k)})}$$

$$\times \int d\mathbf{p} \, W(\mathbf{p})\delta\left(-\frac{(\hbar k)^2}{2m} + \hbar\frac{\mathbf{k} \cdot \mathbf{p}}{m} - \hbar\omega(k)\right)$$

$$- \frac{1}{(2\pi)^2} \int d\mathbf{k} \, \frac{\mathcal{L}^2(k)\hbar\mathbf{k}}{2\omega(k)(e^{\beta\hbar\omega(k)} - 1)} \int d\mathbf{p} \, W(\mathbf{p})\delta\left(\frac{(\hbar k)^2}{2m} + \hbar\frac{\mathbf{k} \cdot \mathbf{p}}{m} - \hbar\omega(k)\right).$$

$$(3.37)$$

Here, according to the notations of Chapter 1,

$$\mathbf{E} = -e_c\mathcal{E}, \quad (3.38)$$

where $\mathcal{E}$ is for the external electric field.

Thus the relation (3.37) is a rigorous consequence of the Boltzmann equation. We assume the "coarse approximation", choosing for of $W(\mathbf{p})$ a "shifted" Maxwellian distribution function with average velocity $\mathcal{V}$,

$$W(\mathbf{p}) = \rho_M(\mathbf{p} - m\mathcal{V}), \quad \rho_M(\mathbf{p}) = \left(\frac{\beta}{2m\pi}\right)^{3/2} e^{-\beta\frac{p^2}{2m}},$$

and substitute this distribution into (3.37). This leads us to the approximate equation

$$e_c \boldsymbol{\mathcal{E}} = \frac{1}{(2\pi)^2} \int d\mathbf{k}\, \frac{\mathcal{L}^2(k)\hbar\mathbf{k}}{2\omega(k)(1 - e^{-\beta\hbar\omega(k)})}$$

$$\times \int d\mathbf{p}\, \rho_M(\mathbf{p})\delta\left(-\frac{(\hbar k)^2}{2m} + \hbar\frac{\mathbf{k}\cdot\mathbf{p}}{m} - \hbar[\omega(k) - \mathbf{k}\cdot\boldsymbol{\mathcal{V}}]\right)$$

$$- \frac{1}{(2\pi)^2} \int d\mathbf{k}\, \frac{\mathcal{L}^2(k)\hbar\mathbf{k}}{2\omega(k)(e^{\beta\hbar\omega(k)} - 1)}$$

$$\times \int d\mathbf{p}\, \rho_M(\mathbf{p})\delta\left(\frac{(\hbar k)^2}{2m} + \hbar\frac{\mathbf{k}\cdot\mathbf{p}}{m} - \hbar[\omega(k) - \mathbf{k}\cdot\boldsymbol{\mathcal{V}}]\right). \quad (3.39)$$

Note that

$$\delta\left(-\frac{(\hbar k)^2}{2m} + \hbar\frac{\mathbf{k}\cdot\mathbf{p}}{m} - \hbar[\omega(k) - \mathbf{k}\cdot\boldsymbol{\mathcal{V}}]\right)$$

$$= \frac{1}{(2\pi)} \int_{-\infty}^{+\infty} \exp\left[i\left(\frac{(\hbar k)^2}{2m} - \hbar\frac{\mathbf{k}\cdot\mathbf{p}}{m} + \hbar[\omega(k) - \mathbf{k}\cdot\boldsymbol{\mathcal{V}}]\right)\xi\right]\, d\xi,$$

$$\delta\left(\frac{(\hbar k)^2}{2m} + \hbar\frac{\mathbf{k}\cdot\mathbf{p}}{m} - \hbar(\omega(k) - \mathbf{k}\cdot\boldsymbol{\mathcal{V}})\right)$$

$$= \frac{1}{(2\pi)} \int_{-\infty}^{+\infty} \exp\left[i\left(\frac{(\hbar k)^2}{2m} + \hbar\frac{\mathbf{k}\cdot\mathbf{p}}{m} - \hbar[\omega(k) - \mathbf{k}\cdot\boldsymbol{\mathcal{V}}]\right)\xi\right]\, d\xi$$

and

$$\int \rho_M(\mathbf{p})e^{-i\xi\frac{\hbar\mathbf{k}\cdot\mathbf{p}}{m}}\, d\mathbf{p} = \exp\left(-\frac{(\hbar k)^2}{2m}\frac{\xi^2}{\beta}\right),$$

$$\int \rho_M(\mathbf{p})e^{i\xi\frac{\hbar\mathbf{k}\cdot\mathbf{p}}{m}}\, d\mathbf{p} = \exp\left(-\frac{(\hbar k)^2}{2m}\frac{\xi^2}{\beta}\right).$$

Therefore, we find from (3.39) that

$$e_c \boldsymbol{\mathcal{E}} = \int_{-\infty}^{+\infty} d\xi\, \frac{1}{(2\pi)^3} \int d\mathbf{k}\, \frac{\mathcal{L}^2(k)\hbar\mathbf{k}}{2\omega(k)}$$

$$\times \left(\frac{e^{i\hbar[\omega(k) - \mathbf{k}\cdot\boldsymbol{\mathcal{V}}]\xi}}{1 - e^{-\beta\hbar\omega(k)}} - \frac{e^{-i\hbar[\omega(k) - \mathbf{k}\cdot\boldsymbol{\mathcal{V}}]\xi}}{e^{\beta\hbar\omega(k)} - 1}\right)\exp\left[-\frac{(\hbar k)^2}{2m}\left(\frac{\xi^2}{\beta} - i\xi\right)\right]. \quad (3.40)$$

This approximate equation was first derived by K. K. Thornber and R. P. Feynman [30] for weak interactions.[1] They found that the mobility derived from (3.40) in the weak-interaction limit does not coincide with the mobility found by the standard method from the Boltzmann equation. We see here that this discrepancy originates from an inadequate approximation, namely the choice of a Maxwellian distribution concentrated in the "vicinity of the average velocity" $\mathcal{V}$, as a trial momentum distribution function in (3.37). This distribution is itself the rigorous consequence of some Boltzmann equation. The connection between (2.13) and the use of a Maxwellian for the trial equilibrium distribution function was also noticed by J. T. Devreese (private communication).

3.3. Nonequilibrium Properties of the Linear Polaron Model

In this section we are going to show that the results of [6] and [30] regarding the impedance calculations in the polaron model can be derived immediately without functional integration.

Let us begin with the rigorous equation (3.36), in which we choose

$$f(\mathbf{p}) = \mathbf{p}. \tag{3.41}$$

for the arbitrary function $f(\mathbf{p})$. We denote the average electron momentum by

$$\langle \mathbf{p} \rangle_t = \int \mathbf{p}\, W_t(\mathbf{p})\, d\mathbf{p}.$$

Introducing notation

$$\mathrm{Tr}_{(S,\Sigma)}\, e^{i\mathbf{k}\cdot\mathbf{r}(\tau)} e^{-i\mathbf{k}\cdot\mathbf{r}(t)} D_{t_0} = \Phi_k(t,\tau,t_0), \tag{3.42}$$

and, in addition,

$$\mathrm{Tr}_{(S,\Sigma)}\, e^{i\mathbf{k}\cdot\mathbf{r}(t)} e^{-i\mathbf{k}\cdot\mathbf{r}(\tau)} D_{t_0} = \mathrm{Tr}_{(S,\Sigma)} \{ e^{i\mathbf{k}\cdot\mathbf{r}(\tau)} e^{-i\mathbf{k}\cdot\mathbf{r}(t)} \}^\dagger D_{t_0} = \Phi_k^*(t,\tau,t_0).$$

it follows from (3.36), on taking (3.41) and (3.42) into account, that

$$\frac{d\langle \mathbf{p}\rangle_t}{dt} + e^{\varepsilon t}\mathbf{E}(t) = -\frac{1}{V}\, e^{2\varepsilon t}\sum_{(k)} \frac{\mathcal{L}^2(k)\mathbf{k}}{2\omega(k)(1 - e^{-\beta\hbar\omega(k)})}$$

$$\times \int_{t_0}^{t} d\tau\, [e^{i\omega(k)(t-\tau)} + e^{-i\omega(k)(t-\tau)}e^{-\beta\hbar\omega(k)}]e^{-\varepsilon(t-\tau)}\Phi_k(t,\tau,t_0)$$

---

[1] According to the notation and system of units used in [30],

$$\hbar = 1, \quad C_k = \frac{1}{V^{1/2}}\left(\frac{1}{2\omega(k)}\right)^{1/2}\mathcal{L}(k), \quad \mathbf{E} = e_c\boldsymbol{\mathcal{E}},$$

equation (3.40) takes the form

$$\mathbf{E} = \int_{-\infty}^{+\infty} d\xi \sum_{(k)} |C_k|^2 \left(\frac{e^{i\hbar[\omega(k) - \mathbf{k}\cdot\boldsymbol{\mathcal{V}}]\xi}}{1 - e^{-\beta\hbar\omega(k)}} - \frac{e^{-i\hbar[\omega(k) - \mathbf{k}\cdot\boldsymbol{\mathcal{V}}]\xi}}{e^{\beta\hbar\omega(k)} - 1} \right) \exp\left[-\frac{k^2}{2m}\left(\frac{\xi^2}{\beta} - i\xi\right) \right].$$

This equation corresponds to formula (3.17) of [30].

$$- \frac{1}{V} e^{2\varepsilon t} \sum_{(k)} \frac{\mathcal{L}^2(\mathbf{k})\mathbf{k}}{2\omega(k)(1 - e^{-\beta\hbar\omega(k)})}$$

$$\times \int_{t_0}^{t} d\tau \, e^{-\varepsilon(t-\tau)} [e^{-i\omega(k)(t-\tau)} + e^{i\omega(k)(t-\tau)} e^{-\beta\hbar\omega(k)}] \Phi_k^*(t, \tau, t_0). \quad (3.43)$$

This is still a rigorous relation. To derive some kind of approximate equation, we have to find some approximation for $\Phi_k(t, \tau, t_0)$ in explicit form. To solve this task, it is appropriate to employ a model Hamiltonian that leads to exactly solvable equations of motion. To get the desired approximation, this Hamiltonian should be constructed in such a way that the behavior of the approximate trajectory $\mathbf{r}(t)$ has some resemblance with the exact trajectory given by the Hamiltonian (3.5).

Let us start with the case of zero external field,

$$\mathbf{E} = 0, \quad (3.44)$$

and consider the Hamiltonian

$$H^{(L)} = \frac{p^2}{2m^*} + \frac{C^2 r^2}{2} + \sum_{(k)} \hbar\nu(k) b_k^\dagger b_k$$

$$+ \frac{i}{V^{1/2}} \sum_{(k)} \left(\frac{\hbar}{2\nu(k)} \right)^{1/2} \Lambda(\mathbf{k})\mathbf{k} \cdot \mathbf{r}(b_k + b_{-k}^\dagger), \quad (3.45)$$

where $\Lambda(\mathbf{k})$ is spherically symmetric function of $\mathbf{k}$, $\nu(\mathbf{k})$ is a spherically symmetric function that is strictly positive:

$$\nu(\mathbf{k}) > 0.$$

Until we take the limit $V \to \infty$, we assume that the volume V is finite and the number of terms n_V in all sums over $\mathbf{k}$ is also finite. Then the corresponding Heisenberg equations of motion form a finite linear system of ordinary differential equations with constant coefficients that is exactly solvable in principle, i.e.

$$\frac{d\mathbf{r}}{dt} = \frac{\mathbf{p}}{m},$$

$$\frac{d\mathbf{p}}{dt} = -C^2 \mathbf{r}(t) - \frac{i}{V^{1/2}} \sum_{(k)} \left(\frac{\hbar}{2\nu(k)} \right)^{1/2} \Lambda(k)\mathbf{k}[b_k(t) + b_{-k}^\dagger(t)],$$

$$\frac{b_k(t)}{dt} = -i\nu(k) b_k(t) - \frac{1}{V^{1/2}} \left(\frac{1}{2\hbar\nu(k)} \right)^{1/2} \Lambda(k)\mathbf{k} \cdot \mathbf{r}(t),$$

$$\frac{b_{-k}^\dagger(t)}{dt} = i\nu(k) b_{-k}^\dagger(t) + \frac{1}{V^{1/2}} \left(\frac{1}{2\hbar\nu(k)} \right)^{1/2} \Lambda(k)\mathbf{k} \cdot \mathbf{r}(t), \quad (3.46)$$

$$\mathbf{r}(t_0) = \mathbf{r}, \quad \mathbf{p}(t_0) = \mathbf{p}, \quad b_k(t_0) = b_k, \quad b_{-k}^\dagger(t_0) = b_{-k}^\dagger.$$

Let us show now that the Hamiltonian (3.45) is translationally invariant under a proper choice of the constant C^2. We begin with the identity

$$\sum_{(k)} \hbar\nu(k) \left(b_k^\dagger + \frac{i\mathbf{k}\cdot\mathbf{r}}{V^{1/2}} \frac{\Lambda(k)}{\nu(k)[2\hbar\nu(k)]^{1/2}} \right) \left(b_k - \frac{i\mathbf{k}\cdot\mathbf{r}}{V^{1/2}} \frac{\Lambda(k)}{\nu(k)[2\hbar\nu(k)]^{1/2}} \right)$$

$$= \sum_{(k)} \hbar\nu(k) b_k^\dagger b_k + \frac{i}{\sqrt{V}} \sum_{(k)} \left(\frac{\hbar}{2\nu(k)} \right)^{1/2} \Lambda(k)\mathbf{k}\cdot\mathbf{r}\, b_k$$

$$- \frac{i}{V^{1/2}} \sum_{(k)} \left(\frac{\hbar}{2\nu(k)} \right)^{1/2} \Lambda(k)\mathbf{k}\cdot\mathbf{r}\, b_k^\dagger + \frac{1}{V} \sum_{(k)} \frac{\Lambda^2(k)}{2\nu^2(k)} (\mathbf{k}\cdot\mathbf{r})^2,$$

and note that, thanks to the spherical symmetry of the functions $\Lambda(k)$, $\nu(k)$

$$\frac{1}{V} \sum_{(k)} \frac{\Lambda^2(k)}{\nu^2(k)} (\mathbf{k}\cdot\mathbf{r})^2 = r^2 \frac{1}{V} \sum_{(k)} \frac{\Lambda^2(k)}{3\nu^2(k)} k^2.$$

On account of this,

$$H^{(L)} = \frac{p^2}{2m^*} + \left(C^2 - \frac{1}{V} \sum_{(k)} \frac{\Lambda^2(k)}{3\nu^2(k)} k^2 \right) \frac{r^2}{2}$$

$$+ \sum_{(k)} \hbar\nu(k) \left(b_k^\dagger + \frac{i\mathbf{k}\cdot\mathbf{r}}{V^{1/2}} \frac{\Lambda(k)}{\nu(k)[2\hbar\nu(k)]^{1/2}} \right) \left(b_k - \frac{i\mathbf{k}\cdot\mathbf{r}}{V^{1/2}} \frac{\Lambda(k)}{\nu(k)[2\hbar\nu(k)]^{1/2}} \right).$$

Therefore if we choose

$$C^2 = \frac{1}{V} \sum_{(k)} \frac{\Lambda^2(k)}{3\nu^2(k)} k^2 \tag{3.47}$$

then the Hamiltonian $H^{(L)}$ becomes invariant with respect to the group of translations

$$\mathbf{r} \to \mathbf{r} + \mathbf{R}, \qquad b_k \to b_k + \frac{i\mathbf{k}\cdot\mathbf{R}}{V^{1/2}} \frac{\Lambda(k)}{\nu(k)[2\hbar\nu(k)]^{1/2}}. \tag{3.48}$$

This invariance leads to the existence of a conservation law for a vector $\mathcal{P}$,

$$\frac{d\mathcal{P}}{dt} = 0, \tag{3.49}$$

that could be interpreted as the conservation of a kind of "total momentum". To find an explicit expression for $\mathcal{P}$ we use (3.6), from which it follows that

$$\frac{d[b_k(t) - b^\dagger_{-k}(t)]}{dt} = -i\nu(k)[b_k(t) - b^\dagger_{-k}(t)]$$

$$-2\frac{1}{V^{1/2}}\sum_{(k)}\left(\frac{1}{2\hbar\nu(k)}\right)^{1/2}\Lambda(k)\mathbf{k}\cdot\mathbf{r}(t),$$

and from here

$$\frac{d}{dt}\frac{1}{V^{1/2}}\left(\frac{\hbar}{2\nu(k)}\right)^{1/2}\Lambda(k)\frac{b_k(t) - b^\dagger_{-k}(t)}{\nu(k)}\mathbf{k}$$

$$= -i\frac{1}{V^{1/2}}\sum_{(k)}\left(\frac{\hbar}{2\nu(k)}\right)^{1/2}\Lambda(k)\mathbf{k}[b_k(t) + b^\dagger_{-k}(t)] - \frac{1}{V}\sum_{(k)}\frac{\Lambda^2(k)}{\nu^2(k)}\mathbf{k}\cdot\mathbf{r}(t)\mathbf{k}$$

$$= \frac{d\mathbf{p}(t)}{dt} + C^2\mathbf{r}(t) - \frac{1}{V}\sum_{(k)}\frac{\Lambda^2(k)}{\nu^2(k)}\mathbf{k}\cdot\mathbf{r}(t)\mathbf{k}.$$

But, thanks to (3.47),

$$-\frac{1}{V}\sum_{(k)}\frac{\Lambda^2(k)}{\nu^2(k)}\mathbf{k}\cdot\mathbf{r}(t)\mathbf{k} = \mathbf{r}(t)\frac{1}{V}\sum_{(k)}\frac{\Lambda^2(k)}{3\nu^2(k)}k^2 = C^2\mathbf{r}(t).$$

Therefore

$$\frac{d}{dt}\left[\mathbf{p}(t) - \frac{1}{V^{1/2}}\sum_{(k)}\frac{\mathbf{k}\Lambda(k)}{\nu(k)}\left(\frac{\hbar}{2\nu(k)}\right)^{1/2}[b_k(t) - b^\dagger_{-k}(t)]\right] = 0.$$

It follows from the last equation that the constant "total momentum" vector has the form

$$\mathcal{P} = \mathbf{p} - \frac{1}{V^{1/2}}\sum_{(k)}\frac{\mathbf{k}\Lambda(k)}{\nu(k)}\left(\frac{\hbar}{2\nu(k)}\right)^{1/2}[b_k(t) - b^\dagger_{-k}(t)]. \qquad (3.50)$$

Now let us introduce an external electric field, replacing the Hamiltonian $H^{(L)}$ with

$$\widetilde{H}^{(L)} = H^{(L)} + \mathbf{E}(r)\cdot\mathbf{r}. \qquad (3.51)$$

Because $H^{(L)}$ commutes with $\mathcal{P}$ and because

$$\mathcal{P}_\beta, r_\gamma] = [p_\beta, r_\gamma] = -i\hbar\delta_{\beta\gamma}, \qquad \beta, \gamma = 1, 2, 3,$$

we see that

$$\frac{d\mathcal{P}}{dt} = -\mathbf{E}(t). \tag{3.52}$$

It can be seen that for the Hamiltonian (3.5), under the condition (3.44), the translation group is defined by the transformations

$$\mathbf{r} \to \mathbf{r} + \mathbf{R}, \quad b_k \to b_k e^{-i\mathbf{k}\cdot\mathbf{R}}. \tag{3.53}$$

Under these circumstances, the "total momentum" is determined by the expression

$$\mathcal{P} = \mathbf{p} + \sum_{(k)} \hbar\mathbf{k} b_k^\dagger b_k. \tag{3.54}$$

In the case when the external field is turned on, $\mathcal{P}$ also satisfies (3.52).

Consider the Hamiltonian (3.51), the corresponding Heisenberg equations and the corresponding initial conditions for the statistical operator D_t. We shall use the same form for these initial conditions as in (3.10):

$$D_{t_0} = \rho(S)D_L(\Sigma).$$

But now we put as a natural choice

$$D_\Sigma = \text{const} \cdot \exp\left(-\beta\sum_{(k)} \hbar\nu(k) b_k^\dagger b_k\right).$$

Here D_Σ is the statistical operator for the statistically equilibrium model system Σ. The equations of motion for the whole model system $S + \Sigma$ are

$$m^* \frac{\mathbf{r}(t)}{dt} = \mathbf{p}(t),$$

$$\frac{d\mathbf{p}(t)}{dt} = -C^2\mathbf{r}(t) - \frac{i}{V^{1/2}} \sum_{(k)} \left(\frac{\hbar}{2\nu(k)}\right)^{1/2} \Lambda(k)\mathbf{k}[b_k(t) + b_{-k}^\dagger(t)] - \mathbf{E}(t), \tag{3.55}$$

$$\frac{b_k(t)}{dt} = -i\nu(k)b_k(t) - \frac{1}{V^{1/2}}\left(\frac{1}{2\hbar\nu(k)}\right)^{1/2} \Lambda(k)\mathbf{k}\cdot\mathbf{r}(t),$$

$$\frac{b_{-k}^\dagger(t)}{dt} = i\nu(k)b_{-k}^\dagger + \frac{1}{V^{1/2}}\left(\frac{1}{2\hbar\nu(k)}\right)^{1/2} \Lambda(k)\mathbf{k}\cdot\mathbf{r}(t), \tag{3.56}$$

$$\mathbf{r}(t_0) = \mathbf{r}, \quad \mathbf{p}(t_0) = \mathbf{p}, \quad b_k(t_0) = b_k, \quad b_{-k}^\dagger(t_0) = b_{-k}^\dagger,$$

from which it follows that

$$b_k(t) = b_k e^{-i\nu(k)(t-t_0)} - \frac{1}{V^{1/2}}\left(\frac{1}{2\hbar\nu(k)}\right)^{1/2} \Lambda(k)\int_{t_0}^{t} d\tau\, e^{-i\nu(k)(t-\tau)}\mathbf{k}\cdot\mathbf{r}(\tau),$$

$$b_{-k}^\dagger(t) = b_{-k}^\dagger e^{i\nu(k)(t-t_0)} + \frac{1}{V^{1/2}}\left(\frac{1}{2\hbar\nu(k)}\right)^{1/2} \Lambda(k)\int_{t_0}^{t} d\tau\, e^{i\nu(k)(t-\tau)}\mathbf{k}\cdot\mathbf{r}(\tau).$$

Substitution of these expressions into (3.55) leads to the equation for the electron momentum:

$$\frac{d\mathbf{p}(t)}{dt} + C^2 \mathbf{r}(t) + \frac{i}{V} \sum_{(k)} \frac{\Lambda^2(k)\mathbf{k}}{2\nu(k)} \int_{t_0}^{t} d\tau\, \mathbf{k} \cdot \mathbf{r}(\tau)(e^{i\nu(k)(t-\tau)} - e^{-i\nu(k)(t-\tau)})$$

$$= -\frac{i}{V^{1/2}} \left(\sum_{(k)} \frac{\hbar}{2\nu(k)} \right)^{1/2} \Lambda(k)\mathbf{k}(b_k e^{-i\nu(k)(t-t_0)} + b^\dagger_{-k} e^{i\nu(k)(t-t_0)}) - \mathbf{E}(t).$$

Integrating by parts,

$$i\int_{t_0}^{t} d\tau\, \mathbf{k} \cdot \mathbf{r}(\tau)(e^{i\nu(k)(t-\tau)} - e^{-i\nu(k)(t-\tau)})$$

$$= -\frac{1}{\nu(k)} \int_{t_0}^{t} d\tau\, \mathbf{k} \cdot \mathbf{r}(\tau) \frac{d}{d\tau} \left(e^{i\nu(k)(t-\tau)} + e^{-i\nu(k)(t-\tau)} \right)$$

$$= -2\frac{\mathbf{k} \cdot \mathbf{r}(t)}{\nu(k)} + \frac{2\mathbf{k} \cdot \mathbf{r}}{\nu(k)} \cos[\nu(k)(t-t_0)]$$

$$+ \frac{2}{\nu(k)} \int_{t_0}^{t} d\tau\, \mathbf{k} \cdot \frac{d\mathbf{r}(\tau)}{d\tau} \cos[\nu(k)(t-\tau)],$$

and remembering that

$$-\frac{1}{V} \sum_{(k)} \frac{\Lambda^2(k)}{\nu^2(k)} \mathbf{k} \cdot \mathbf{r}(t)\mathbf{k} = -\frac{1}{V} \sum_{(k)} \frac{\Lambda^2(k)}{3\nu^2(k)} k^2 \mathbf{r}(t) = -C^2 \mathbf{r}(t)$$

$$= \frac{1}{V} \sum_{(k)} \frac{\Lambda^2(k)\mathbf{k}}{\nu^2(k)} \mathbf{k} \cdot \frac{d\mathbf{r}(\tau)}{d\tau} \cos[\nu(k)(t-\tau)]$$

$$= \frac{1}{V} \sum_{(k)} \frac{\Lambda^2(k)k^2}{3\nu^2(k)} \cos[\nu(k)(t-\tau)] \frac{d\mathbf{r}(\tau)}{d\tau},$$

we obtain

$$\frac{d\mathbf{p}(t)}{dt} + \frac{1}{m^*} \int_{t_0}^{t} d\tau\, K(t-\tau)\mathbf{p}(\tau) = -\mathbf{r}K(t-t_0)$$

$$- \frac{i}{V^{1/2}} \left(\frac{\hbar}{2\nu(k)} \right)^{1/2} \Lambda(k)\mathbf{k}(b_k e^{-i\nu(k)(t-t_0)} + b^\dagger_{-k} e^{i\nu(k)(t-t_0)}) - \mathbf{E}(t),$$

$$\tag{3.57}$$

where

$$K(t-\tau) = \frac{1}{V} \sum_{(k)} \frac{\Lambda^2(k)k^2}{3\nu^2(k)} \cos[\nu(k)(t-\tau)]. \tag{3.58}$$

We average this equation with the initial statistical operator

$$D_{t_0} = \rho(S)D(\Sigma) \tag{3.59}$$

and denote

$$m^* \langle \mathcal{V}(t) \rangle = \langle \mathbf{p}(t) \rangle = \text{Tr}_{(S,\Sigma)} \, \mathbf{p}(t) D_{t_0},$$

$$\langle \mathbf{r} \rangle = \text{Tr}_{(S,\Sigma)} \, \mathbf{r} D_{t_0} = \text{Tr}_{(S)} \, \mathbf{r} \rho(S).$$

Because $\langle b_k \rangle = 0$, $\langle b^\dagger_{-k} \rangle = 0$, equation (3.57) can be reduced to

$$m^* \frac{d \langle \mathcal{V}(t) \rangle}{dt} + \int_{t_0}^{t} d\tau \, \langle \mathcal{V}(\tau) \rangle K(t - \tau) = -\langle \mathbf{r} \rangle K(t - t_0) - \mathbf{E}(t). \tag{3.60}$$

Here $\langle \mathcal{V}(t) \rangle$ is the average velocity of the particle.

Let us now consider the situation where $\mathbf{E}(t)$ is a periodic function of t multiplied by $e^{\varepsilon t}$ $(\varepsilon > 0)$, corresponding to the adiabatic switching on of the external electric field at time $t \to -\infty$. We shall seek the stationary solutions of (3.60), i.e. solutions that can be represented as a product of $e^{\varepsilon t}$ and some periodic function.

Since (3.60) is a linear equation, we can restrict ourselves to consideration of the simplest ansatz

$$\mathbf{E}(t) = \mathbf{E}_\omega e^{(-i\omega + \varepsilon)t}. \tag{3.61}$$

In fact, if $\mathbf{E}(t)$ were a sum of terms with different frequencies ω then the resulting stable solutions of (3.60) would be a sum of solution of the type (3.61).

Thus consider the equation

$$m^* \frac{d \langle \mathcal{V}(t) \rangle}{dt} + \int_{-\infty}^{t} d\tau \, \langle \mathcal{V}(\tau) \rangle K(t - \tau) = -\mathbf{E}_\omega e^{(-i\omega + \varepsilon)t}.$$

Substituting

$$\langle \mathcal{V}(t) \rangle = \mathcal{V}_\omega e^{-(i\omega + \varepsilon)t},$$

into this equation, we get

$$\left(m^*(-i\omega + \varepsilon) + \int_0^\infty K(t) e^{(i\omega - \varepsilon)t} \, dt \right) \mathcal{V}_\omega = -\mathbf{E}_\omega.$$

The definition (3.58) leads to the relation

$$\int_0^\infty K(t) e^{(-i\omega + \varepsilon)t} \, dt = \frac{1}{V} \sum_{(k)} \frac{\Lambda^2(k)k^2}{6\nu^2(k)} \left(\frac{1}{\varepsilon - i[\omega + \nu(k)]} + \frac{1}{\varepsilon - i[\omega - \nu(k)]} \right).$$

Denote

$$\frac{1}{V}\sum_{(k)}\frac{\Lambda^2(k)k^2}{6\nu^2(k)}\left[\delta(\nu(k)-\Omega)+\delta(\nu(k)+\Omega)\right]=I(\Omega). \tag{3.62}$$

Then $I(-\Omega)=I(\Omega)$, $I(\Omega)\geqslant 0$, and

$$\int_0^\infty K(t)e^{(i\omega-\varepsilon)t}\,dt=i\int_{-\infty}^{+\infty}I(\Omega)\frac{d\Omega}{\omega+i\varepsilon-\Omega}. \tag{3.63}$$

Therefore

$$\left(m^*(-i\omega+\varepsilon)+i\int_{-\infty}^{+\infty}I(\Omega)\frac{d\Omega}{\omega+i\varepsilon-\Omega}\right)\langle\boldsymbol{\mathcal{V}}(t)\rangle=-\mathbf{E}_\omega e^{(-i\omega+\varepsilon)t}.$$

But, thanks to (3.38),

$$\mathbf{E}_\omega=-e_c\boldsymbol{\mathcal{E}}_\omega,$$

and, according to the definition of the electric current,

$$\mathbf{j}_\omega(t)=-e_0\langle\boldsymbol{\mathcal{V}}(t)\rangle,$$

we have

$$\left(m^*(-i\omega+\varepsilon)+i\int_{-\infty}^{+\infty}I(\Omega)\frac{d\Omega}{\omega+i\varepsilon-\Omega}\right)\mathbf{j}_\omega(t)=e_c^2\boldsymbol{\mathcal{E}}_\omega e^{(-i\omega+\varepsilon)t}. \tag{3.64}$$

Let us take the limit $V\to\infty$, assuming that for any real ω and any positive ε,

$$\int_{-\infty}^{+\infty}I(\Omega)\frac{d\Omega}{\omega+i\varepsilon-\Omega}\to\int_{-\infty}^{+\infty}J(\Omega)\frac{d\Omega}{\omega+i\varepsilon-\Omega}. \tag{3.65}$$

After the passage to the limit, we put $\varepsilon\to 0$ in (3.64). We get

$$\mathbf{j}_\omega(t)=\frac{1}{Z_+(\omega)}e_c^2\boldsymbol{\mathcal{E}}_\omega e^{-i\omega t},$$

where

$$Z_+(\omega)=-m^*i\omega+i\int_{-\infty}^{+\infty}J(\Omega)\frac{d\Omega}{\omega-\Omega+i0}. \tag{3.66}$$

Choosing a system of units for which the electron charge e_c is unity, we see that (3.66) is exactly the impedance corresponding to the frequency $-\omega$.

As we shall see later in connection with the process of the passage to the limit, all expressions used further, including (3.42), will depend only on the function $J(\Omega)$,, but not on the particular choice of functions $\nu(k)$ and $\Lambda(k)$. Therefore we have to employ, first of all, an appropriate expression for $J(\Omega)$. Let us allot the following properties to this function:

1) $J(\Omega)$ is an analytic function of the complex variable, regular on the strip

$$|\operatorname{Im}\Omega| \leqslant \eta_0.$$

2) $J(\Omega) = J(-\Omega)$.
3) $|J(\Omega)| \leqslant C/|\Omega|^2$ for $|\Omega| \geqslant \omega_0$, where ω_0, C are constants.
4) For real Ω

$$J(\Omega) > 0. \tag{3.67a}$$

Then we take expressions for $\Lambda(k)$ and $\nu(k)$ such that [m]

$$\frac{1}{V} \sum_{\nu(k) \geqslant \omega} \frac{\Lambda^2(k)k^2}{6\nu^2(k)} < \frac{C_1}{\omega}, \quad \text{where } C_1 \text{ is some constant independent of } V,$$

$$\tag{3.67b}$$

$$\frac{1}{V} \sum_{\nu(k) \leqslant \omega} \frac{\Lambda^2(k)k^2}{6\nu^2(k)} \to \int_0^\infty J(\Omega)\, d\Omega, \quad 0 < \omega < \infty. \tag{3.67c}$$

In the considered situation it is clear that (3.65) holds for any fixed $\varepsilon > 0$ and that the convergence is uniform with respect to ω in the interval $-\infty < \omega < +\infty$).

Let us now introduce the following function of the complex variable W:

$$\Delta(W) = i \int_{-\infty}^{+\infty} J(\Omega) \frac{d\Omega}{W - \Omega}. \tag{3.68}$$

We see that this function is regular for $|\operatorname{Im} W| > 0$. In connection with the properties (3.67), it is obvious that

$$\Delta(W) = \lim_{V \to \infty} i \int_{-\infty}^{+\infty} I(\Omega) \frac{d\Omega}{W - \Omega}, \quad \operatorname{Im} W \neq 0. \tag{3.69}$$

Here, thanks to (3.62),

$$i \int_{-\infty}^{+\infty} I(\Omega) \frac{d\Omega}{W - \Omega} = \frac{i}{V} \sum_{(k)} \frac{\Lambda^2(k)k^2}{6\nu^2(k)} \left(\frac{1}{W - \nu(k)} + \frac{1}{W + \nu(k)} \right),$$

and hence this function is analytic on the whole complex plane and has singularities (poles) only on the real axis at $W = \pm\nu(k)$. However, the limit function has a cut along the whole real axis, such that

$$\Delta(\omega + i0) - \Delta(\omega - i0) = 2\pi J(\omega) > 0.$$

---

[m] One of the ways to find such expressions for the functions $\Lambda(k)$ and $\nu(k)$ results in the following. We take $\mathbf{k} = \left(\dfrac{2\pi n_1}{L}, \dfrac{2\pi n_2}{L}, \dfrac{2\pi n_3}{L} \right)$, $L^3 = V$; n_1, n_2, n_3 to be positive and negative integers; it is assumed that $n_1^2 + n_2^2 + n_3^2 \neq 0$, this prevents the appearance of zero value for $\mathbf{k}$ in all sums over $\mathbf{k}$. Then we put $\nu(k) = s|\mathbf{k}|$, $\Lambda(k) = 2\pi^2(s^3/|\mathbf{k}|^2)J(s|\mathbf{k}|)$, where s is some positive constant independent of V.

Thus we have two analytic functions

$$\Delta_+(W) = i \int\limits_{-\infty}^{+\infty} J(\Omega)\frac{d\Omega}{W - \Omega} \quad \text{for} \quad \text{Im}\, W \geqslant 0,$$

$$\Delta_-(W) = -i \int\limits_{-\infty}^{+\infty} J(\Omega)\frac{d\Omega}{W - \Omega} \quad \text{for} \quad \text{Im}\, W \leqslant 0.$$

(3.70)

Thanks to properties (2) and (4) in (3.67a), these functions are connected with each other in a simple way:

$$\Delta_-(W) = -\Delta_+(-W) \quad \text{for} \quad \text{Im}\, W < 0. \tag{3.71}$$

Hence, we need to investigate only one of them, for example, $\Delta_+(-W)$. Denote

$$\text{Re}\, W = \omega, \qquad \text{Im}\, W = y > 0. \tag{3.72}$$

Then, for any fixed $\omega_1 > 0$

$$\Delta_+(\omega + iy) = i \int\limits_{|\Omega - \omega| > \omega_1} J(\Omega)\frac{d\Omega}{\omega + iy - \Omega} + i \int\limits_{\omega - \omega_1}^{\omega + \omega_1} J(\Omega)\frac{\omega - \Omega - iy}{(\omega - \Omega)^2 + y^2}\, d\Omega.$$

But

$$\int\limits_{\omega - \omega_1}^{\omega + \omega_1} \frac{\omega - \Omega}{(\omega - \Omega)^2 + y^2}\, d\Omega = -\int\limits_{-\omega_1}^{\omega_1} \frac{\Omega}{\Omega^2 + y^2}\, d\Omega = 0,$$

and therefore

$$\Delta_+(\omega + iy) = i \int\limits_{|\Omega - \omega| > \omega_1} J(\Omega)\frac{d\Omega}{\omega + iy - \Omega}$$

$$+ i \int\limits_{\omega - \omega_1}^{\omega + \omega_1} \frac{J(\Omega) - J(\omega)}{(\omega - \Omega)^2 + y^2}(\omega - \Omega)\, d\Omega + \int\limits_{\omega - \omega_1}^{\omega + \omega_1} J(\Omega)\frac{y}{(\omega - \Omega)^2 + y^2}\, d\Omega. \tag{3.73}$$

From which it follows that

$$\Delta_+(\omega) = \lim_{y \to 0} \Delta_+(\omega + iy)$$

$$= i \int\limits_{|\Omega - \omega| > \omega_1} J(\Omega)\frac{d\Omega}{\omega - \Omega} + i \int\limits_{\omega - \omega_1}^{\omega + \omega_1} \frac{J(\Omega) - J(\omega)}{\omega - \Omega}\, d\Omega + \pi J(\omega). \tag{3.74}$$

Thus $\Delta_+(\omega)$ is also an analytic function on the real axis. Using (3.73), it is easy to prove that

$$|\Delta_+(\omega)| \leqslant \frac{\text{const}}{|W|}, \qquad |W| \to \infty. \tag{3.75}$$

Furthermore, we have

$$\Delta_+(\omega) = \Delta_-(\omega) + 2\pi J(\omega) = -\Delta_+(-\omega) + 2\pi J(\omega).$$

Thanks to condition (1), the function $-\Delta_+(-\Omega) + 2\pi J(\omega)$ is analytic in the domain

$$0 \geqslant \operatorname{Im} W \geqslant -\eta_0. \tag{3.76}$$

Because this function coincides with $\Delta_+(\omega)$ on the real axis, we see that $\Delta_+(\omega)$, defined initially for $\operatorname{Im} W > 0$, can be continued analytically to the domain (3.76). Thus we can write

$$\Delta_+(W) = -\Delta_+(-W) + 2\pi J(W) \quad \text{for} \quad 0 \geqslant \operatorname{Im} W \geqslant -\eta_0. \tag{3.77}$$

It can be shown that the inequality (3.75) is justified anywhere for

$$\operatorname{Im} W \geqslant -\eta_0. \tag{3.78}$$

Let us consider now the impedance function

$$Z_+(W) = -im^* W + \Delta_+(\Omega)$$

in the domain (3.78) and note that it does not have any zeros in the upper half-plane or on the real axis, because, thanks to (3.73),

$$\operatorname{Re} Z_+(W) > 0 \quad \text{for} \quad \operatorname{Im} W \geqslant 0.$$

Hence all zeros of this function in the considered domain (3.78), if any, must be confined within the domain (3.76) . But

$$\Delta_+(W) \to 0 \quad \text{for} \quad |W| \to \infty,$$

and therefore zeros of the function $Z_+(W)$ might be observed only in the closed domain

$$|\operatorname{Re} W| \leqslant \text{const}, \quad 0 \geqslant \operatorname{Im} W \geqslant -\eta_0. \tag{3.79}$$

As is well known, an analytic function can possess only a finite number of zeros in any closed domain. If a few zeros are contained in the domain (3.70) then we can choose $\eta > 0$ such that $-\eta$ is larger than any of the imaginary parts of these zero points. If, on the other hand, the domain (3.79) does not contain any zeros of the function $Z_+(W)$ then we choose $\eta = \eta_0$. In any case, we see that by choosing an appropriate value $\eta > 0$, we can always ensure that the domain

$$\operatorname{Im} W \geqslant -\eta \tag{3.80}$$

does not contain any zeros of the impedance function $Z_+(W)$. Therefore the admittance function $1/Z_+(W)$ is a regular analytic function in the domain (3.80). Its behavior at infinity is given by the relation

$$\frac{1}{Z_+(W)} = \frac{1}{-m^* iW + \Delta^+(W)} = -\frac{1}{m^* iW} + \frac{\Delta^+(W)}{m^* iW(-m^* iW + \Delta^+(W))}$$

$$= -\frac{1}{m^* iW} + O\left(\frac{1}{W^3}\right), \quad |W| \to \infty. \tag{3.81}$$

In conclusion, we should like to consider the following example. Let us take

$$\Delta^+(W) = i\frac{K_0^2}{2}\left(\frac{1}{W - \nu_0 + i\gamma} + \frac{1}{W + \nu_0 + i\gamma}\right), \quad \gamma > 0, \quad \text{Im}\, W > -\gamma,$$

$$\Delta^-(W) = -\Delta^+(-W) = i\frac{K_0^2}{2}\left(\frac{1}{W + \nu_0 - i\gamma} + \frac{1}{W - \nu_0 - i\gamma}\right), \quad \text{Im}\, W < \gamma.$$

Then

$$J(\omega) = \frac{1}{(2\pi)}\left[\Delta^+(\omega) - \Delta^-(\omega)\right] = \frac{K_0^2}{2\pi}\left(\frac{\gamma}{(\omega - \nu_0)^2 + \gamma^2} + \frac{\gamma}{(\omega + \nu_0)^2 + \gamma^2}\right).$$

$$(3.82)$$

For this example, all of our conditions are fulfilled. A similar result would be obtained if, instead of the single term in (3.82), the sum of a few terms of this type were considered.

After these lengthy speculations on the analyticity of the impedance and admittance functions, we return to our fundamental equation (3.57), in which we put

$$\mathbf{E}(t) = \sum_\omega \boldsymbol{\mathcal{E}}_\omega e^{-i\omega t}. \tag{3.83}$$

It is convenient to solve this equation by the Laplace transform. Thus we multiply both sides of the equation by the factor

$$e^{iWt}, \quad W = \Omega + i\delta, \tag{3.84}$$

and integrate over t:

$$\int_{t_0}^\infty dt\, e^{iWt}\frac{d\mathbf{p}(t)}{dt} + \frac{1}{m^*}\int_{t_0}^\infty e^{iWt}\int_{t_0}^t d\tau\, K(t-\tau)\mathbf{p}(\tau)$$

$$= -\mathbf{r}\int_{t_0}^\infty dt\, e^{iWt}K(t - t_0) - \sum_\omega \boldsymbol{\mathcal{E}}_\omega \int_{t_0}^\infty e^{i(W-\omega)t}\, dt - \frac{i}{V^{1/2}}\left(\frac{\hbar}{2\nu(k)}\right)^{1/2}\Lambda(k)\mathbf{k}$$

$$\times \left(b_k\int_{t_0}^\infty dt\, e^{i(W-\nu(k))t}e^{i\nu(k)t_0} + b_{-k}^\dagger\int_{t_0}^\infty dt\, e^{i(W+\nu(k))t}e^{-i\nu(k)t_0}\right). \tag{3.85}$$

But

$$\int_{t_0}^{\infty} dt\, e^{iWt} \frac{d\mathbf{p}(t)}{dt} = -\mathbf{p}e^{iWt_0} - iW \int_{t_0}^{\infty} dt\, e^{iWt}\mathbf{p}(t),$$

$$\int_{t_0}^{\infty} dt\, e^{iWt} \int_{t_0}^{t} d\tau\, K(t-\tau)\mathbf{p}(\tau) - \int_{0}^{\infty} K(t)e^{iWt}\, dt \int_{t_0}^{\infty} e^{iWt}\mathbf{p}(t)\, dt.$$

Therefore

$$\frac{1}{m^*}\left(-im^*W + \int_{0}^{\infty} K(t)e^{iWt}\, dt\right) \int_{t_0}^{\infty} e^{iWt}\mathbf{p}(t)\, dt$$

$$= \mathbf{p}e^{iWt_0} - \mathbf{r}e^{iWt_0} \int_{0}^{\infty} K(t)e^{iWt}\, dt + \sum_{(\omega)} \mathbf{E}_\omega \frac{e^{i(W-\omega)t_0}}{i(W-\omega)}$$

$$+ \frac{1}{V^{1/2}} \sum_{(k)} \left(\frac{\hbar}{2\nu(k)}\right)^{1/2} \Lambda(k)\mathbf{k}\left(\frac{b_k e^{iWt_0}}{W-\nu(k)} + \frac{b_{-k}^\dagger e^{iWt_0}}{W+\nu(k)}\right).$$

From here, thanks to (3.63), we have

$$\int_{0}^{\infty} K(t)e^{iWt}\, dt = i \int_{-\infty}^{+\infty} I(\nu)\frac{d\nu}{W-\nu}.$$

Introducing the notation

$$-im^*W + i \int_{-\infty}^{+\infty} I(\nu)\frac{d\nu}{W-\nu} = Z^{(V)}(W), \qquad i \int_{-\infty}^{+\infty} I(\nu)\frac{d\nu}{W-\nu} = \Delta^V(W). \tag{3.86}$$

we get

$$\int_{t_0}^{\infty} dt\, e^{iWt}\mathbf{p}(t)$$

$$= \frac{m^*\mathbf{p}e^{iWt_0}}{Z^{(V)}(W)} - m^*\mathbf{r}\frac{\Delta^{(V)}(W)}{Z^{(V)}(W)}e^{iWt_0} - i\sum_{(\omega)} m^*\mathbf{E}_\omega \frac{e^{iWt_0}e^{-i\omega t_0}}{(W-\omega)Z^{(V)}(W)}$$

$$+ \frac{1}{V^{1/2}} \sum_{(k)} m^* \left(\frac{\hbar}{2\nu(k)}\right)^{1/2} \frac{\Lambda(k)\mathbf{k}e^{iWt_0}}{Z^{(V)}(W)}\left(\frac{b_k}{W-\nu(k)} + \frac{b_{-k}^\dagger}{W+\nu(k)}\right). \tag{3.87}$$

Because

$$f(t) = \frac{1}{(2\pi)} \int_{-\infty}^{+\infty} e^{(\delta-i\Omega)t}\left(\int_{t_0}^{\infty} f(\tau)e^{(i\Omega-\delta)\tau}\, d\tau\right)\, d\Omega, \quad t > t_0,$$

Therefore, using the notation

$$\frac{i}{2\pi} \int_{-\infty}^{+\infty} \frac{e^{(\delta-i\Omega)(t-t_0)}}{(\Omega+i\delta-\nu)Z^{(V)}(\Omega+i\delta)}\, d\Omega = f(\nu, \delta, t-t_0),$$

$$\frac{1}{2\pi}\int_{-\infty}^{+\infty}\frac{e^{(\delta-i\Omega)(t-t_0)}}{Z^{(V)}(\Omega+i\delta)}\,d\Omega=\frac{1}{2\pi m^*}\int_{-\infty}^{+\infty}\left(-\frac{1}{i\Omega-\delta}+\frac{\Delta^{(V)}(\Omega+i\delta)}{(i\Omega-\delta)Z^{(V)}(\Omega+i\delta)}\right)$$

$$\times\,e^{(\delta-i\Omega)(t-t_0)}\,d\Omega=g_0(\delta,t-t_0),$$

$$\frac{1}{2\pi}\int_{-\infty}^{+\infty}\frac{\Delta^{(V)}(\Omega+i\delta)}{Z^{(V)}(\Omega+i\delta)}\,e^{(\delta-i\Omega)(t-t_0)}\,d\Omega=g_1(\delta,t-t_0),\qquad(3.88)$$

we obtain from (3.87)

$$\mathbf{p}(t)=\mathbf{p}^{(S)}(t)+\mathbf{p}^{(E)}(t)+\mathbf{p}^{(\Sigma)}(t),$$

$$\mathbf{p}^{(S)}(t)=m^*\mathbf{p}g_0(\delta,t-t_0)-m^*\mathbf{r}g_1(\delta,t-t_0),\qquad(3.89)$$

$$\mathbf{p}^{(E)}(t)=-m^*\sum_{(\omega)}\mathbf{E}_\omega f(\omega,\delta,t-t_0)e^{-i\omega t},$$

$$\mathbf{p}^{(\Sigma)}(t)=\frac{-im^*}{V^{1/2}}\sum_{(k)}\left(\frac{\hbar}{2\nu(k)}\right)^{1/2}$$

$$\times\,\Lambda(k)\mathbf{k}[b_k f(\nu(k),\delta,t-t_0)+b_{-k}^\dagger f(-\nu(k),\delta,t-t_0)].$$

It should be stressed that the functions (3.88) depend essentially on V.

Thanks to our choice, which leads to conditions1)–4) and (3.67), we can take the passage to the limit $V\to\infty$. Up to the present, δ has been arbitrary. Let us now choose

$$\delta=\frac{\eta}{2}.\qquad(3.90)$$

From the other side, it is easy to see that

$$g_0(\delta,t-t_0)\to\Psi_0(t-t_0)=\frac{1}{2\pi}\int_{-\infty}^{+\infty}\frac{e^{(\delta-i\Omega)(t-t_0)}}{Z_+(\Omega+i\delta)}\,d\Omega,$$

$$g_1(\delta,t-t_0)\to\Psi_1(t-t_0)=\frac{1}{2\pi}\int_{-\infty}^{+\infty}\frac{\Delta_+(\Omega+i\delta)}{Z_+(\Omega+i\delta)}\,e^{(\delta-i\Omega)(t-t_0)}\,d\Omega,\quad(3.91)$$

$$f(\nu,\delta,t-t_0)\to\Phi(\nu,t-t_0)=\frac{i}{2\pi}\int_{-\infty}^{+\infty}\frac{e^{(\delta-i\Omega)(t-t_0)}\,d\Omega}{(\Omega+i\delta-\nu)Z_+(\Omega+i\delta)}$$

as $V\to\infty$. Taking into account the identity

$$\frac{1}{Z(\Omega+i\delta)}=\frac{1}{\delta-i\Omega}+\frac{\Delta(\Omega+i\delta)}{(i\Omega-\delta)Z(\Omega+i\delta)}$$

and the magnitude δ fixed by (3.90), one can show that the convergence

$$|f(\nu, \delta, t - t_0) - \Phi(\nu, t - t_0)| \to 0,$$
$$\nu |f(\nu, \delta, t - t_0) - \Phi(\nu, t - t_0)| \to 0, \qquad (3.92)$$
$$V \to \infty$$

is uniform with respect to real ν when $|t - t_0| \leqslant T$. Here T is some constant independent on V.

Let us begin to study the asymptotic behavior of the limiting functions Ψ_0, Ψ_1 and Φ for $t - t_0 \to \infty$. It is suitable to remind that functions $\Delta_+(\Omega + i\delta)$; $1/Z_+(\Omega + i\delta)$ are regular analytic functions of Ω in the domain $\operatorname{Im}\Omega \geqslant -\delta - \eta = -3\delta$. Therefore the integration in the expressions for Ψ_0 and Ψ_1 can be distorted from the real axis to the axis $(-3i\delta - \infty, -3i\delta + \infty)$, which implies the change of variables

$$\Omega \to \Omega - 3i\delta.$$

Because of this,

$$\Psi_1(t - t_0) = \frac{1}{2\pi} \int_{-\infty}^{+\infty} \frac{\Delta_+(\Omega - i\eta)}{Z_+(\Omega - i\eta)} e^{-i\Omega(t-t_0)} \, d\Omega \, e^{-\eta(t-t_0)},$$

$$\Psi_0(t - t_0) = \frac{1}{2\pi} \int_{-\infty}^{+\infty} \frac{e^{-i\Omega(t-t_0)}}{Z_+(\Omega - i\eta)} \, d\Omega \, e^{-\eta(t-t_0)}$$
$$= \frac{1}{2\pi} \int_{-\infty}^{+\infty} \frac{\Delta_+(\Omega - i\eta)}{(\eta + i\Omega)Z_+(\Omega - i\eta)} e^{-i\Omega(t-t_0)} \, d\Omega \, e^{-\eta(t-t_0)},$$

since

$$\int_{-\infty}^{+\infty} \frac{e^{-i\Omega(t-t_0)}}{\eta + i\Omega} \, d\Omega = 0$$

for $t > t_0$. Hence, taking into account the inequalities proved before, we have

$$|\Psi_1(t - t_0)| \leqslant K_1 e^{-\eta(t-t_0)},$$
$$\Psi_0(t - t_0) \leqslant K_0 e^{-\eta(t-t_0)}, \qquad (3.93)$$
$$t > t_0,$$

where K_0 and K_1 are some constants. We apply a similar procedure to the function $\Phi(\nu, t - t_0)$. But here we need only pay attention to the fact that the function under the integral (3.91) has a pole at $\Omega = \nu - i\delta$ in the domain $\operatorname{Im}\Omega + \delta \geqslant -\eta$. As a consequence,

$$\Phi(\nu, t - t_0) = \frac{e^{-i\nu(t-t_0)}}{Z_+(\nu)} + \frac{i}{2\pi} e^{-\eta(t-t_0)} \int_{-\infty}^{+\infty} \frac{e^{-i\Omega(t-t_0)} \, d\Omega}{(\Omega + i\delta - \nu)Z_+(\Omega - i\eta)}$$

$$= \frac{e^{-i\nu(t-t_0)}}{Z_+(\nu)} + \frac{i}{2\pi} e^{-\eta(t-t_0)} \int_{-\infty}^{+\infty} \frac{\Delta_+(\Omega - i\eta)e^{-i\Omega(t-t_0)} \, d\Omega}{(\Omega - i\eta - \nu)(i\Omega + \eta)Z_+(\Omega - i\eta)}, \qquad (3.94)$$

because

$$\int_{-\infty}^{+\infty} \frac{e^{-i\Omega(t-t_0)}}{(\Omega - i\eta - \nu)(i\Omega - i\eta)} \, d\Omega = 0, \quad t > t_0.$$

The expressions (3.94) lead to the following inequalities:

$$\left| \Phi(\nu, t - t_0) - \frac{e^{-i\nu(t-t_0)}}{Z_+(\nu)} \right| \leqslant K_2 e^{-\eta(t-t_0)},$$

$$\left| \nu\Phi(\nu, t - t_0) - \nu\frac{e^{-i\nu(t-t_0)}}{Z_+(\nu)} \right| \leqslant K_3 e^{-\eta(t-t_0)}.$$

(3.95)

Here K_2 and K_3 are some constants.

Now we can pass to the calculation of the expressions (3.42) for the model based on the Hamiltonian $H^{(L)}$. We have

$$\Phi_k^{(a)}(t, \tau, t_0) = \text{Tr} \, e^{i\mathbf{k}\cdot\mathbf{r}(\tau)} e^{-i\mathbf{k}\cdot\mathbf{r}(t)} D_{t_0}.$$

(3.96)

Here the index (a) indicates that we have used an approximation: instead of the function $\mathbf{r}(t)$ determined by the exact equations of motion, the function $\mathbf{r}(t)$ given by (3.56), which follows from the model Hamiltonian $H^{(L)}$, has been substituted.

The approximate equation, which we propose to solve instead of the exact one (3.43), is formulated as

$$\frac{\langle \mathbf{p} \rangle_t}{dt} + \mathbf{E}(t) = - \lim_{\varepsilon>0,\varepsilon\to 0} \frac{1}{(2\pi)^3} \int d\mathbf{k} \, \frac{\mathbf{k}\mathcal{L}^2(k)}{2\omega(k)(1 - e^{-\beta\hbar\omega(k)})} \int_{-\infty}^{t} d\tau \, e^{-\varepsilon(t-\tau)}$$

$$\times [e^{i\omega(k)(t-\tau)} + e^{-i\omega(k)(t-\tau)} e^{-\beta\hbar\omega(k)}] \lim_{t_0\to-\infty} \lim_{V\to\infty} \Phi_k^{(a)}(t, \tau, t_0)$$

$$- \lim_{\varepsilon>0,\varepsilon\to 0} \frac{1}{(2\pi)^3} \int d\mathbf{k} \, \frac{\mathbf{k}\mathcal{L}^2(k)}{2\omega(k)(1 - e^{-\beta\hbar\omega(k)})} \int_{-\infty}^{t} d\tau \, e^{-\varepsilon(t-\tau)}$$

$$\times [e^{-i\omega(k)(t-\tau)} + e^{i\omega(k)(t-\tau)} e^{-\beta\hbar\omega(k)}] \lim_{t_0\to-\infty} \lim_{V\to\infty} \Phi_k^{*(a)}(t, \tau, t_0). \quad (3.97)$$

We note that this equation follows from (3.43) by replacing Φ_k with $\Phi_k^{(a)}$ and taking the sequence of limits $V \to \infty$, $t_0 \to -\infty$, $\varepsilon \to 0$.

To write the approximate equation (3.97) in explicit form, we consider the expression (3.96) for $\Phi_k^{(a)}$. Let us pay attention first of all to the fact that the "model equations" (3.56) are linear, so the commutators

$$[\mathbf{r}_j(t), \mathbf{r}_{j'}(t)], \quad j, j' = 1, 2, 3$$

are c-numbers. Therefore

$$e^{i\mathbf{k}\cdot\mathbf{r}(\tau)}e^{-i\mathbf{k}\cdot\mathbf{r}(t)} = e^{\frac{1}{2}[\mathbf{k}\cdot\mathbf{r}(\tau),\mathbf{k}\cdot\mathbf{r}(t)]}e^{i\mathbf{k}\cdot[\mathbf{r}(t)-\mathbf{r}(\tau)]},$$

$$\Phi_k^{(a)}(t,\tau,t_0) = e^{-i\mathbf{k}\cdot\mathbf{r}(t)} = e^{\frac{1}{2}[\mathbf{k}\cdot\mathbf{r}(\tau),\mathbf{k}\cdot\mathbf{r}(t)]}\mathrm{Tr}_{(S,\Sigma)}\exp\left(-i\mathbf{k}\int_\tau^t \frac{\mathbf{p}(s)}{m^*}\,ds\right)D_{t_0}.$$

$$(3.98)$$

Inserting (3.89) into this equation and observing that

$$D_{t_0} = \rho(S)d(\Sigma), \quad D(\Sigma) = \mathrm{const}\cdot\exp\left(-\beta\sum_{(k)}\hbar\nu(k)b_k^+ b_k\right), \quad (3.99)$$

$$\mathrm{Tr}_{(S)}\,\rho(S) = 1, \quad \mathrm{Tr}_{(\Sigma)}\,D(\Sigma) = 1,$$

we get from (3.98)

$$\Phi_k^{(a)}(t,\tau,t_0) = \Phi_k^{(1)}(t,\tau,t_0)\Phi_k^{(2)}(t,\tau,t_0), \quad (3.100)$$

$$\Phi_k^{(1)}(t,\tau,t_0) = e^{\frac{1}{2}[\mathbf{k}\cdot\mathbf{r}(\tau),\mathbf{k}\cdot\mathbf{r}(t)]}\exp\left(-i\frac{\mathbf{k}}{m^*}\cdot\int_\tau^t \mathbf{p}^{(E)}(s)\,ds\right)$$

$$\times\,\mathrm{Tr}_{(\Sigma)}\exp\left(-i\frac{\mathbf{k}}{m^*}\cdot\int_\tau^t \mathbf{p}^{(\Sigma)}(s)\,ds\right)D(\Sigma), \quad (3.101)$$

$$\Phi_k^{(2)}(t,\tau,t_0) = \mathrm{Tr}_{(S)}\exp\left(-i\frac{\mathbf{k}}{m^*}\cdot\int_\tau^t \mathbf{p}^{(S)}(s)\,ds\right)\rho(S). \quad (3.102)$$

Furthermore, we have

$$[\mathbf{k}\cdot\mathbf{r}(\tau),\mathbf{k}\cdot\mathbf{r}(t)] = \frac{1}{m^*}\left[\mathbf{k}\cdot\mathbf{r}(\tau),\int_\tau^t \mathbf{k}\cdot\mathbf{p}(s)\,ds\right], \quad \mathbf{r}(\tau) = \mathbf{r}(t) - \int_\tau^t \frac{\mathbf{p}(s)}{m}\,ds,$$

$$[\mathbf{k}\cdot\mathbf{r}(\tau),\mathbf{k}\cdot\mathbf{p}(s)] = [\mathbf{k}\cdot\{\mathbf{r}(\tau)-\mathbf{r}(s)\},\mathbf{k}\cdot\mathbf{p}(s)] + i\hbar k^2$$

$$= -\frac{1}{m^*}\left[\int_\tau^s \mathbf{k}\cdot\mathbf{p}(\sigma)\,d\sigma,\mathbf{k}\cdot\mathbf{p}(s)\right] + i\hbar k^2.$$

Thus

$$[\mathbf{k}\cdot\mathbf{r}(\tau),\mathbf{k}\cdot\mathbf{r}(t)] = \frac{i\hbar k^2}{m^*}(t-\tau) - \left(\frac{1}{m^*}\right)^2\int_\tau^t ds\int_\tau^s d\sigma\,[\mathbf{k}\cdot\mathbf{p}(\sigma),\mathbf{k}\cdot\mathbf{p}(s)]$$

$$= i\hbar\frac{k^2}{m^*}(t-\tau) + \left(\frac{1}{m^*}\right)^2\int_\tau^t ds\int_s^t d\sigma[\mathbf{k}\cdot\mathbf{p}(\sigma),\mathbf{k}\cdot\mathbf{p}(s)],$$

because

$$\int\limits_{\tau}^{t} ds \int\limits_{\tau}^{t} d\sigma \, [\mathbf{k} \cdot \mathbf{p}(\sigma), \mathbf{k} \cdot \mathbf{p}(s)] = \left[\int\limits_{\tau}^{t} \mathbf{k} \cdot \mathbf{p}(\xi) \, d\xi, \int\limits_{\tau}^{t} \mathbf{k} \cdot \mathbf{p}(\xi) \, d\xi \right] = 0$$

and

$$\int\limits_{\tau}^{t}\int\limits_{\tau}^{s} A + \int\limits_{\tau}^{t}\int\limits_{s}^{t} A = \int\limits_{\tau}^{t}\int\limits_{\tau}^{t} A = 0.$$

On the basis of (3.89) and observing that all terms in the sum

$$\mathbf{p}(t) = \mathbf{p}^{(S)}(t) + \mathbf{p}^{(\Sigma)}(t) + \mathbf{p}^{(E)}(t)$$

commute with each other, we find that

$$[\mathbf{k} \cdot \mathbf{r}(\tau), \mathbf{k} \cdot \mathbf{r}(t)] = \frac{i\hbar k^2}{m^*}(t - \tau) + \left(\frac{1}{m^*}\right)^2 \int\limits_{\tau}^{t} ds \int\limits_{s}^{t} d\sigma [\mathbf{k} \cdot \mathbf{p}^{(S)}(\sigma), \mathbf{k} \cdot \mathbf{p}^{(S)}(s)]$$

$$+ \left(\frac{1}{m^*}\right)^2 \int\limits_{\tau}^{t} ds \int\limits_{s}^{t} d\sigma [\mathbf{k} \cdot \mathbf{p}^{(\Sigma)}(\sigma), \mathbf{k} \cdot \mathbf{p}^{(\Sigma)}(s)]. \quad (3.103)$$

Here, because of (3.89),

$$\left(\frac{1}{m^*}\right)^2 [\mathbf{k} \cdot \mathbf{p}^{(S)}(\sigma), \mathbf{k} \cdot \mathbf{p}^{(S)}(s)]$$

$$= i\hbar k^2 [g_0(\delta, \sigma - t_0) g_1(\delta, s - t_0) - g_0(\delta, s - t_0) g_1(\delta, \sigma - t_0)] \quad (3.104)$$

and

$$\left(\frac{1}{m^*}\right)^2 [\mathbf{k} \cdot \mathbf{p}^{(\Sigma)}(\sigma), \mathbf{k} \cdot \mathbf{p}^{(\Sigma)}(s)] = k^2 [F(\sigma, s, t_0) - F(s, \sigma, t_0)], \quad (3.105)$$

where

$$F(\sigma, s, t_0) = \frac{1}{V} \sum_{(k)} \frac{\hbar}{6\nu(k)} \frac{\Lambda^2(k) k^2}{(1 - e^{-\beta\hbar\nu(k)})} [f(\nu(k), \delta, \sigma - t_0)$$

$$\times f(-\nu(k), \delta, s - t_0) + e^{-\beta\hbar\nu(k)} f(-\nu(k), \delta, \sigma - t_0) f(\nu(k), \delta, s - t_0)]$$

or

$$F(\sigma, s, t_0) = \int\limits_{0}^{\infty} d\nu \, I(\nu) \frac{\hbar\nu}{1 - e^{-\beta\hbar\nu}} [f(\nu(k), \delta, \sigma - t_0) f(-\nu(k), \delta, s - t_0)$$

$$+ e^{-\beta\hbar\nu(k)} f(-\nu(k), \delta, \sigma - t_0) f(\nu(k), \delta, s - t_0)]. \quad (3.106)$$

Let us note further that because of the form (3.99) of the operator $D_L(\Sigma)$ and the linearity of $\mathbf{p}^{(\Sigma(s))}$ with respect to the Bose operators, we can write [n]:

$$\mathrm{Tr}_{(\Sigma)} \exp\left(-i\frac{\mathbf{k}}{m^*} \cdot \int_\tau^t \mathbf{p}^{(\Sigma)}(s)\,ds\right) D_L(\Sigma)$$

$$= \exp\left[-\frac{1}{2m^2}\mathrm{Tr}_{(\Sigma)}\left(\int_\tau^t \mathbf{p}^{(\Sigma)}(s)\,ds\right)^2 D_L(\Sigma)\right]$$

$$= \exp\left(-\frac{1}{2m^2}\int_\tau^t ds \int_\tau^t d\sigma\, \mathrm{Tr}_{(\Sigma)}\, \mathbf{k}\cdot\mathbf{p}^{(\Sigma)}(s)\,\mathbf{k}\cdot\mathbf{p}^{(\Sigma)}(\sigma)D_L(\Sigma)\right)$$

$$= \exp\left(-\frac{k^2}{2}\int_\tau^t ds \int_\tau^t d\sigma\, F(s,\sigma,t_0)\right). \tag{3.107}$$

Let us also recall that

$$\frac{1}{m^*}\mathbf{p}^E(t) = -\sum_{(\omega)}\mathbf{E}_\omega f(\omega,\delta,t-t_0)e^{-i\omega t_0}. \tag{3.108}$$

Bearing in mind (3.67), (3.91) and (3.92), we get

$$[\mathbf{k}\cdot\mathbf{r}(\tau),\mathbf{k}\cdot\mathbf{r}(t)]_{V\to\infty} \to i\hbar\frac{k^2}{m^*}(t-\tau)$$

$$+ i\hbar k^2 \int_\tau^t ds \int_s^t d\sigma\, [\Psi_0(\sigma-t_0)\Psi_1(s-t_0) - \Psi_0(s-t_0)\Psi_1(\sigma-t_0)]$$

$$+ k^2 \int_\tau^t ds \int_s^t d\sigma [F_\infty(\sigma,s,t_0) - F_\infty(s,\sigma,t_0)]. \tag{3.109}$$

Here, in connection with (3.106),

$$F_\infty(\sigma,s,t_0) = \lim_{V\to\infty} F(\sigma,s,t_0) = \int_0^\infty J(\nu)\frac{\hbar\nu}{1-e^{-\beta\hbar\nu}}$$

$$\times [\Phi(\nu,\sigma-t_0)\Phi(-\nu,s-t_0) + e^{-\beta\hbar\nu}\Phi(\nu,s-t_0)\Phi(-\nu,\sigma-t_0)]\,d\nu.$$

$$\tag{3.110}$$

We also have, from (3.108),

$$\frac{1}{m^*}\mathbf{p}^{(E)}(t) \to -\sum_{(\omega)}\mathbf{E}_\omega\Phi(\omega,t-t_0)e^{-i\omega t_0}. \tag{3.111}$$

---

[n] Here $\langle e^A \rangle = e^{\langle A^2 \rangle/2}$, where A is a linear form in Bose operators, has been used, and the averaging is with respect to the quadratic Hamiltonian $\sum_{(\omega)} E_\omega b_\omega^\dagger b_\omega$.

Then we can write (see (3.100))

$$\Phi_k^{(1)}(t, \tau, t_0) \to e^{\frac{1}{2} \lim\limits_{V \to \infty} [\mathbf{k} \cdot \mathbf{r}(\tau), \, \mathbf{k} \cdot \mathbf{r}(t)]} \exp\left(-i \frac{\mathbf{k}}{m^*} \int\limits_\tau^t \lim\limits_{V \to \infty} \mathbf{p}^{(E)}(s)\, ds\right)$$

$$\times \exp\left(-\frac{k^2}{2} \int\limits_\tau^t ds \int\limits_\tau^t d\sigma \, F_\infty(s, \sigma, t_0)\right), \quad (3.112)$$

taking into account (3.109), (3.100) and (3.111).

Now, let us consider the situation when $t_0 \to -\infty$. Because of (3.95) and (3.111), we observe that

$$\left| \lim\limits_{V \to \infty} \frac{1}{m^*} \mathbf{p}^{(E)}(t) - \mathbf{v}(t) \right| \leqslant \sum\limits_{(\omega)} |\mathbf{E}_\omega| K_2 e^{-\eta(t - t_0)}, \quad (3.113)$$

$$\mathbf{v}(t) = -\sum\limits_{(\omega)} \frac{\mathbf{E}_\omega}{Z_+(\omega)} e^{-i\omega t}. \quad (3.114)$$

Furthermore, incorporating (3.95) for the evaluation of the expression (3.100), we find

$$|F_\infty(\sigma, s, t_0) - F(\sigma - s)| \leqslant \widetilde{K}(e^{-\eta(\sigma - t_0)} + e^{-\eta(s - t_0)}), \quad \widetilde{K} = \text{const}, \quad (3.115)$$

$$F(\sigma - s) = \int\limits_0^\infty d\nu \, J(\nu) \frac{\hbar\nu}{1 - e^{-\beta\hbar\nu}} \frac{e^{-i\nu(\sigma - s)} + e^{-\beta\hbar\nu} e^{i\nu(\sigma - s)}}{Z_+(\nu) Z_+(-\nu)},$$

or, as long as $J(\nu) = J(-\nu)$,

$$F(\sigma - s) = \int\limits_{-\infty}^{+\infty} d\nu \, J(\nu) \frac{\hbar\nu}{1 - e^{-\beta\hbar\nu}} \frac{e^{-i\nu(\sigma - s)}}{Z_+(\nu) Z_+(-\nu)}. \quad (3.116)$$

We transform this formula slightly. Because

$$2\pi J(\nu) = \Delta_+(\nu) - \Delta_-(\nu) = Z_+(\nu) - Z_-(\nu) = Z_+(\nu) + Z_+(-\nu),$$

we have

$$\frac{1}{Z_+(\nu) Z_+(-\nu)} J(\nu) = \frac{1}{2\pi} \frac{Z_+(\nu) + Z_+(-\nu)}{Z_+(\nu) Z_+(-\nu)} = \frac{1}{2\pi} \left(\frac{1}{Z_+(\nu)} + \frac{1}{Z_+(-\nu)} \right)$$

$$= \frac{1}{2\pi} \left(\frac{1}{Z_+(\nu)} - \frac{1}{Z_-(\nu)} \right).$$

Furthermore, because of the reality of $J(\nu)$, the following relation holds

$$\text{Im}\, Z_+(\nu) = -\text{Im}\, Z_+(-\nu).$$

But, by definition,

$$Z_+(\nu) = i \int\limits_{-\infty}^{+\infty} \frac{J(\Omega)}{\Omega - \nu + i0}\, d\Omega,$$

and hence
$$\operatorname{Re} Z_+(\nu) = \pi J(\nu) = \operatorname{Re} Z_+(-\nu),$$

so that
$$Z_+(-\nu) = Z_+^*(\nu),$$

As a consequence, we may rewrite (3.116) in the form

$$F(\sigma - s) = \int\limits_{-\infty}^{+\infty} G(\nu) \frac{\hbar\nu}{1 - e^{-\beta\hbar\nu}} e^{-i\nu(\sigma - s)} \, d\nu,$$

$$G(\nu) = \frac{1}{2\pi} \left(\frac{1}{Z_+(\nu)} - \frac{1}{Z_+(\nu)} \right) = \frac{J(\nu)}{|Z_-(\nu)|^2}.$$

(3.117)

As the final result, keeping in mind (3.101), (3.109), (3.112), (3.115) and (3.93), we obtain

$$\lim_{t_0 \to -\infty} \lim_{V \to \infty} \Phi_k^{(1)}(t, \tau, t_0) = e^{-i\int\limits_{\tau}^{t} \mathbf{k}\cdot\boldsymbol{\mathcal{V}}(s))\,ds} A(k^2, t - \tau),$$

(3.118)

where

$$A(k^2, t - \tau) = \exp\left[k^2\left(\frac{i\hbar}{2m^*}(t - \tau) + \frac{1}{2}\int\limits_{\tau}^{t} ds \int\limits_{s}^{t} d\sigma \left[F(\sigma - s) - F(s - \sigma)\right]\right.\right.$$

$$\left.\left. - \frac{1}{2}\int\limits_{\tau}^{t} ds \int\limits_{\tau}^{t} d\sigma \, F(s - \sigma)\right)\right].$$

(3.119)

But, because of (3.117),

$$\frac{1}{2}\int\limits_{\tau}^{t} ds \int\limits_{\tau}^{t} d\sigma \, F(s - \sigma) = \int\limits_{-\infty}^{+\infty} G(\nu) \frac{1 - \cos\nu(t - \tau)}{\nu^2} \frac{\hbar\nu}{1 - e^{-\beta\hbar\nu}} \, d\nu,$$

$$F(\sigma - s) - F(s - \sigma) = 2i \int\limits_{-\infty}^{+\infty} G(\nu) \frac{\hbar\nu}{1 - e^{-\beta\hbar\nu}} \sin\nu(s - \sigma) \, d\nu$$

$$= 2i \int\limits_{-\infty}^{+\infty} G(\nu) \frac{\hbar\nu}{1 - e^{\beta\hbar\nu}} \sin\nu(s - \sigma) \, d\nu$$

$$= -2i \int\limits_{-\infty}^{+\infty} G(\nu) \frac{e^{-\beta\hbar\nu}\hbar\nu}{1 - e^{-\beta\hbar\nu}} \sin\nu(s - \sigma) \, d\nu = \int\limits_{-\infty}^{+\infty} G(\nu)\hbar\nu \sin\nu(s - \sigma) \, d\nu,$$

$$\int\limits_{\tau}^{t} ds \int\limits_{s}^{t} d\sigma \left[F(\sigma - s) - F(s - \sigma)\right] = i\hbar \int\limits_{-\infty}^{+\infty} G(\nu) \left(\frac{\sin\nu(t - \tau)}{\nu} - (t - \tau)\right) \, d\nu,$$

Thus (3.119) yields

$$A(k^2, t, \tau) = A(k^2, t - \tau)$$

$$= \exp\left\{ k^2 \left[\frac{i\hbar}{2m^*}(t - \tau) + \frac{i\hbar}{2} \int\limits_{-\infty}^{+\infty} G(\nu) \left(\frac{\sin \nu(t - \tau)}{\nu} - (t - \tau) \right) d\nu \right. \right.$$

$$\left. \left. - \frac{1}{2} \int\limits_{-\infty}^{+\infty} G(\nu) \frac{\hbar[1 - \cos \nu(t - \tau)]}{\nu(1 - e^{-\beta\hbar\nu})} \, d\nu \right] \right\}. \quad (3.120)$$

Now consider the expression

$$\Phi_k^{(2)}(t, \tau, t_0) = \mathrm{Tr}_{(S)} \exp\left(-\frac{i}{m^*} \int\limits_{\tau}^{t} \mathbf{k} \cdot \mathbf{p}^{(S)}(\sigma) \, d\sigma \right) \rho(S).$$

Here

$$\frac{i}{m^*} \int\limits_{\tau}^{t} \mathbf{k} \cdot \mathbf{p}^{(S)}(\sigma) \, d\sigma = \mathbf{k} \cdot \mathbf{p} \int\limits_{\tau}^{t} g_0(\delta, \sigma - t_0) \, d\sigma - \mathbf{k} \cdot \mathbf{r} \int\limits_{\tau}^{t} g_1(\delta, \sigma - t_0) \, d\sigma.$$

In accordance with the results mentioned above,

$$\int\limits_{\tau}^{t} g_0(\delta, \sigma - t_0) \, d\sigma \to \int\limits_{\tau}^{t} \Psi_0(\sigma - t_0) \, d\sigma,$$

$$\int\limits_{\tau}^{t} g_1(\delta, \sigma - t_0) \, d\sigma \to \int\limits_{\tau}^{t} \Psi_1(\sigma - t_0) \, d\sigma, \quad V \to \infty, \quad (3.121)$$

and also

$$\left| \int\limits_{\tau}^{t} \Psi_0(\sigma - t_0) \, d\sigma \right| \leqslant K_0 \int\limits_{\tau}^{t} e^{-\eta(\sigma - t_0)} \, d\sigma,$$

$$\left| \int\limits_{\tau}^{t} \Psi_1(\sigma - t_0) \, d\sigma \right| \leqslant K_1 \int\limits_{\tau}^{t} e^{-\eta(\sigma - t_0)} \, d\sigma. \quad (3.122)$$

From here it is natural to see that

$$\Phi_k^{(2)}(t, \tau, t_0) = \mathrm{Tr}_{(S)} \rho(S)$$

$$\times \exp\left(-i\mathbf{k} \cdot \mathbf{p} \int\limits_{\tau}^{t} g_0(\delta, \sigma - t_0) \, d\sigma + i\mathbf{k} \cdot \mathbf{r} \int\limits_{\tau}^{t} g_1(\delta, \sigma - t_0) \, d\sigma \right)$$

$$\to \mathrm{Tr}_{(S)} \rho(S) \exp\left(-i\mathbf{k} \cdot \mathbf{p} \int\limits_{\tau}^{t} \Psi_0(\sigma - t_0) \, d\sigma + i\mathbf{k} \cdot \mathbf{r} \int\limits_{\tau}^{t} \Psi_1(\sigma - t_0) \, d\sigma \right)$$

$$\quad (3.123)$$

as $V \to \infty$, and

$$\mathrm{Tr}_{(S)} \rho(S) \exp\left(-i\mathbf{k} \cdot \mathbf{p} \int\limits_{\tau}^{t} \Psi_0(\sigma - t_0) \, d\sigma + i\mathbf{k} \cdot \mathbf{r} \int\limits_{\tau}^{t} \Psi_1(\sigma - t_0) \, d\sigma \right)$$

$$\to \mathrm{Tr}_{(S)} \rho(S) = 1. \quad (3.124)$$

However, in spite of (3.121) and (3.122), considerable difficulties arise in the proof of (3.123) and (3.124). They originate from the unboundedness of the operators $\mathbf{r}$ and $\mathbf{p}$.

Nevertheless, the validity of (3.123) and (3.124) can be confirmed (see Appendix II) in the case when the statistical operator $\rho(S)$ does not depend on V or t_0. In this case,

$$\lim_{t_0 \to -\infty} \lim_{V \to \infty} \Phi_k^{(2)}(t, \tau, t_0) = 1, \tag{3.125}$$

and hence, on the basis of (3.100) and (3.118), we conclude that

$$\lim_{t_0 \to -\infty} \lim_{V \to \infty} \Phi_k^{(a)}(t, \tau, t_0) = e^{-i \int_\tau^t \mathbf{k} \cdot \mathcal{V}(\sigma)) \, d\sigma} A(k^2, t - \tau). \tag{3.126}$$

We substitute this expression into the approximate equation (3.97):

$$\frac{d\langle \mathbf{p} \rangle_t}{dt} + \mathbf{E}(t) = -\lim_{\varepsilon > 0, \varepsilon \to 0} \frac{1}{(2\pi)^3} \int d\mathbf{k} \, \frac{k \mathcal{L}^2(k)}{2\omega(k)(1 - e^{-\beta\hbar\omega(k)})} \int_{-\infty}^t d\tau \, e^{-\varepsilon(t-\tau)}$$

$$\times (e^{i\omega(k)(t-\tau)} + e^{-i\omega(k)(t-\tau)} e^{-\beta\hbar\omega(k)}) e^{-i \int_\tau^t \mathbf{k} \cdot \mathcal{V}(\sigma) \, d\sigma} A(k^2, t - \tau)$$

$$+ \lim_{\varepsilon > 0, \varepsilon \to 0} \frac{1}{(2\pi)^3} \int d\mathbf{k} \, \frac{k \mathcal{L}^2(k)}{2\omega(k)(1 - e^{-\beta\hbar\omega(k)})} \int_{-\infty}^t d\tau \, e^{-\varepsilon(t-\tau)}$$

$$\times (e^{-i\omega(k)(t-\tau)} + e^{i\omega(k)(t-\tau)} e^{-\beta\hbar\omega(k)}) e^{i \int_\tau^t \mathbf{k} \cdot \mathcal{V}(\sigma) \, d\sigma} A^*(k^2, t - \tau). \tag{3.127}$$

Thus we have derived a general approximate equation from which all results of [6, 30] could be deduced. Note that in[6, 30], $m = m^*$ and the function (3.82) is used in the limit $\gamma \to 0$.

Consider, in particular, the case of a weak external field, when one can restrict oneself to a linear approximation of the average velocity with respect to $\mathbf{E}$. Put in (3.127)

$$e^{\pm i \int_\tau^t \mathbf{k} \cdot \mathcal{V}(\sigma)) \, d\sigma} = 1 \pm i \int_\tau^t \mathbf{k} \cdot \mathcal{V}(\sigma) \, d\sigma.$$

Bearing in mind the radial symmetry, we find

$$\frac{\langle \mathbf{p} \rangle_t}{dt} + \mathbf{E}(t) = -\lim_{\varepsilon \to 0, \varepsilon > 0} \frac{i}{(2\pi)^3} \int d\mathbf{k} \, \frac{k^2 \mathcal{L}^2(k)}{6\omega(k)(1 - e^{-\beta\hbar\omega(k)})} \int_{-\infty}^t d\tau \, e^{-\varepsilon(t-\tau)}$$

$$\times (e^{i\omega(k)(t-\tau)} + e^{-i\omega(k)(t-\tau)} e^{-\beta\hbar\omega(k)}) \int_\tau^t \mathcal{V}(\sigma) \, d\sigma \, A(k^2, t - \tau)$$

$$- \lim_{\varepsilon \to 0, \varepsilon > 0} \frac{i}{(2\pi)^3} \int d\mathbf{k} \, \frac{k^2 \mathcal{L}^2(k)}{6\omega(k)(1 - e^{-\beta\hbar\omega(k)})} \int_{-\infty}^t d\tau \, e^{-\varepsilon(t-\tau)}$$

$$\times (e^{-i\omega(k)(t-\tau)} + e^{i\omega(k)(t-\tau)} e^{-\beta\hbar\omega(k)}) \int_\tau^t \mathcal{V}(\sigma) \, d\sigma \, A^*(k^2, t - \tau). \tag{3.128}$$

Here

$$\mathcal{V}(t) = -\sum_{(\omega)} \frac{\mathbf{E}_\omega}{Z_+(\omega)} e^{-i\omega t}$$

represents the stationary average velocity induced in the model system by the external field

$$\mathbf{E}(t) = \sum_{(\omega)} \mathbf{E}_\omega e^{-i\omega t}.$$

Note that in the considered case of weak-enough field, we can express the stationary average velocity for the real system analogously:

$$\frac{\langle \mathbf{p}(t) \rangle}{m} = -\sum_{(\omega)} \frac{\mathbf{E}_\omega}{z_+(\omega)} e^{-i\omega t},$$

but, of course, with different coefficients; here $z_+(\omega)$ corresponds to the impedance of the real system.

To derive a "self-consistent" equation to determine this impedance, we choose

$$Z_+(\omega) = z_+(\omega) \tag{3.129}$$

and take advantage of (3.128). We get

$$-i\int_\tau^t \mathcal{V}(\sigma)\, d\sigma = -\sum_{(\omega)} \frac{\mathbf{E}_\omega e^{-i\omega t}}{Z_+(\omega)} \frac{1 - e^{i\omega(k)(t-\tau)}}{\omega}.$$

Introducing the integration argument $t - \tau = s$, we find

$$m^* \sum_{(\omega)} \frac{\mathbf{E}_\omega i\omega e^{-i\omega t}}{z_+(\omega)} + \sum_{(\omega)} \mathbf{E}_\omega e^{-i\omega t} = \lim_{\varepsilon > 0, \varepsilon \to 0} \sum_{(\omega)} \frac{\mathbf{E}_\omega e^{-i\omega t}}{z_+(\omega)}$$

$$\times \left(\frac{1}{(2\pi)^3} \int d\mathbf{k} \, \frac{k^2 \mathcal{L}^2(k)}{6\omega(k)(1 - e^{-\beta\hbar\omega(k)})} \int_0^\infty ds\, e^{-\varepsilon s} (e^{-i\omega(k)s} + e^{i\omega(k)s} e^{-\beta\hbar\omega(k)}) \right.$$

$$\times \frac{e^{i\omega s} - 1}{\omega} A^*(k^2, s) - \frac{1}{(2\pi)^3} \int d\mathbf{k} \, \frac{k^2 \mathcal{L}^2(k)}{6\omega(k)(1 - e^{-\beta\hbar\omega(k)})}$$

$$\times \left. \int_0^\infty ds\, e^{-\varepsilon s} (e^{i\omega(k)s} + e^{-i\omega(k)s} e^{-\beta\hbar\omega(k)}) \frac{e^{i\omega s} - 1}{\omega} A(k^2, s) \right). \tag{3.130}$$

Thus the impedance is determined self-consistently from the approximate equation

$$z_+(\omega) = -i\omega m + \lim_{\varepsilon > 0, \varepsilon \to 0} \frac{1}{(2\pi)^2} \int d\mathbf{k} \, \frac{k^2 \mathcal{L}^2(k)}{6\omega(k)(1 - e^{-\beta\hbar\omega(k)})}$$

$$\times \int_0^\infty ds\, e^{-\varepsilon s} \left((e^{-i\omega(k)s} + e^{i\omega(k)s} e^{-\beta\hbar\omega(k)}) \frac{e^{i\omega s} - 1}{\omega} A^*(k^2, s) \right.$$

$$\left. - (e^{i\omega(k)s} + e^{-i\omega(k)s} e^{-\beta\hbar\omega(k)}) \frac{e^{i\omega s} - 1}{\omega} A(k^2, s) \right). \tag{3.131}$$

Consider, as an example, (3.127) in the case of a constant field, when

$$\mathbf{E} = \text{const}, \quad \mathcal{V} = \text{const.} \tag{3.132}$$

Since, because of (3.119), $A^*(k^2, s) = A(k^2, -s)$, on making the transformation $\mathbf{k} \to -\mathbf{k}$ in the terms containing $e^{\pm i\omega(k)s}e^{-\beta\hbar\omega(k)}$, we can write:

$$-\mathbf{E} = \lim_{(\varepsilon > 0, \varepsilon \to 0)} \frac{1}{(2\pi)^3} \int d\mathbf{k} \, \frac{\mathbf{k}\mathcal{L}^2(k)}{2\omega(k)} \int_{-\infty}^{+\infty} ds \, e^{-\varepsilon|s|}$$

$$\times \left(\frac{e^{i[\omega(k) - \mathbf{k}\cdot\mathcal{V}]s}}{1 - e^{-\beta\hbar\omega(k)}} - \frac{e^{-i[\omega(k) - \mathbf{k}\cdot\mathcal{V}]s}}{e^{\beta\hbar\omega(k)} - 1} \right) A(k^2, s). \tag{3.133}$$

Equation (3.40) follows from here in the case of weak interaction if we substitute $A(k^2, s)$ with its "zeroth-order approximation" (which neglects interaction) and put $s = \hbar\xi$.

In conclusion, we should make a few remarks on the structure of the stationary probability density for the momentum of the particle S in the model system described by the Hamiltonian $H^{(L)}$. Letting $w_t(\mathbf{p})$, stand for this probability density distribution, we have

$$\int e^{-i\boldsymbol{\lambda}\cdot\mathbf{p}} w_t(\mathbf{p}) \, d\mathbf{p} = \text{Tr}_{(S,\Sigma)} e^{-i\boldsymbol{\lambda}\cdot\mathbf{p}(t)} D_{t_0}$$

$$= e^{-i\boldsymbol{\lambda}\cdot\mathbf{p}^{(E)}(t)} \text{Tr}_{(S)} e^{-i\boldsymbol{\lambda}\cdot\mathbf{p}^{(S)}(t)} \rho(S) \text{Tr}_{(\Sigma)} e^{-i\boldsymbol{\lambda}\cdot\mathbf{p}^{(\Sigma)}(t)} D_L(\Sigma). \tag{3.134}$$

Here, as before,

$$e^{-i\boldsymbol{\lambda}\cdot\mathbf{p}^{(\Sigma)}(t)} D_L(\Sigma) = \exp\left(-\frac{\lambda^2 m^{*2}}{2} F(t, t, t_0) \right). \tag{3.135}$$

We now recall that

$$\mathbf{p}^{(E)}(t) \underset{V\to\infty}{\to} \mathbf{p}_\infty^{(E)}(t) = -m^* \sum_{(\omega)} \mathbf{E}_\omega \Phi(\omega, t - t_0) e^{-i\omega t_0},$$

$$\mathbf{p}_\infty^{(E)}(t) - m^*\mathcal{V}(t) \underset{t\to+\infty}{\to} 0, \tag{3.136}$$

$$\mathbf{p}^{(S)}(t) \underset{V\to\infty}{\to} \mathbf{p}_\infty(t) = m^*\mathbf{p}\Psi_0(t - t_0) - m^*\mathbf{r}\Psi_1(t - t_0)$$

and

$$F(t, t, t_0) \underset{V\to\infty}{\to} F_\infty(t_1, t, t_0), \tag{3.137}$$

$$|F_\infty(t, t, t_0) - F(0)| < 2\widetilde{K} e^{-\eta(t - t_0)} \underset{t\to+\infty}{\to} 0$$

Here

$$F(0) = \int_{-\infty}^{+\infty} \frac{J(\nu)}{|Z_+(\nu)|^2} \frac{\hbar\nu}{1 - e^{-\beta\hbar\nu}} \, d\nu > 0. \tag{3.138}$$

Repeating arguments used to investigate the function $\Phi_k^{(a)}$, we find, as a consequence of (3.134),

$$\int e^{-i\boldsymbol{\lambda}\cdot\mathbf{P}} w_t(\mathbf{p}) \, d\mathbf{p} \underset{V\to\infty}{\longrightarrow} e^{-i\boldsymbol{\lambda}\cdot\mathbf{P}_\infty^{(E)}(t)} \operatorname{Tr}_{(S)} e^{-i\boldsymbol{\lambda}\cdot\mathbf{P}_\infty^{(S)}(t)} \rho(S)$$

$$\times \exp\left(-\frac{\lambda^2 m^{*2}}{2} F_\infty(t,t,t_0) \right) \quad (3.139)$$

and

$$\lim_{V\to\infty} \int e^{-i\boldsymbol{\lambda}\cdot\mathbf{P}} w_t(\mathbf{p}) \, d\mathbf{p} - e^{-im^*\boldsymbol{\lambda}\cdot\mathbf{v}(t)} \exp\left(-\frac{\lambda^2 m^{*2}}{2} F(0) \right) = 0. \quad (3.140)$$

Let us now consider the momentum distribution function itself:

$$\omega_t(\mathbf{p}) = \frac{1}{(2\pi)^3} \int d\boldsymbol{\lambda}\, e^{i\boldsymbol{\lambda}\cdot\mathbf{P}} \left(\int e^{-i\boldsymbol{\lambda}\cdot\mathbf{P}} w_t(\mathbf{p}) \, d\mathbf{p} \right). \quad (3.141)$$

Note that

$$\left| e^{i\boldsymbol{\lambda}\cdot\mathbf{P}} \int e^{-i\boldsymbol{\lambda}\cdot\mathbf{P}} w_t(\mathbf{p}) \, d\mathbf{p} \right| \leqslant \exp\left(-\frac{\lambda^2 m^{*2}}{2} F(t,t,t_0) \right). \quad (3.142)$$

But, thanks to (3.137) and (3.139), it is easy to prove that

$$F_\infty(t,t,t_0) > \frac{F(0)}{2} > 0 \quad (3.143)$$

for a large-enough difference $t - t_0$. Fix such t and t_0. Because, for fixed t and t_0, $F(t,t,t_0) \to F_\infty(t,t,t_0)$ as $V \to \infty$, we see that, for large enough V,

$$F_\infty(t,t,t_0) > \frac{F(0)}{4} > 0$$

and

$$\left| e^{i\boldsymbol{\lambda}\cdot\mathbf{P}} \int e^{-i\boldsymbol{\lambda}\cdot\mathbf{P}} w_t(\mathbf{p}) \, d\mathbf{p} \right| \leqslant e^{\frac{i\lambda^2 m^{*2}}{8} F(0)}.$$

Therefore the passage to the limit in (3.141) as $V \to \infty$ can be carried out on the basis of (3.139) before the integration over $\boldsymbol{\lambda}$. Hence we get

$$\lim_{V\to\infty} \omega_t(\mathbf{p}) = \frac{1}{(2\pi)^3} \int d\boldsymbol{\lambda}\, e^{i\boldsymbol{\lambda}\cdot[\mathbf{p}-\mathbf{P}_\infty^{(E)}(t)]} \left[\operatorname{Tr}_{(S)} e^{-i\boldsymbol{\lambda}\cdot\mathbf{P}_\infty^{(S)}(t)} \rho(S) \right]$$

$$\times \exp\left(-\frac{\lambda^2 m^{*2}}{2} F_\infty(t,t,t_0) \right). \quad (3.144)$$

Thanks to (3.143), the absolute value of the integrand in (3.144) will be smaller than $e^{\frac{i\lambda^2 m^{*2}}{4} F(0)}$ for a large-enough difference $t - t_0$. As a consequence, we can again take the limit $t \to \infty$ under the integral over λ in (3.144) and obtain, in accordance with (3.140),

$$\lim_{t\to\infty}\left[\omega_t(\mathbf{p}) - \frac{1}{(2\pi)^3}\int d\lambda\, \exp\left(i\lambda \cdot [\mathbf{p} - m^*\mathcal{V}(t)] - \frac{\lambda^2 m^{*2}}{2} F(0)\right)\right] = 0.$$
(3.145)

Having calculated the Gaussian integral, we find that

$$\frac{1}{(2\pi)^3}\int d\lambda\, \exp\left(i\lambda \cdot [\mathbf{p} - m^*\mathcal{V}(t)] - \frac{\lambda^2 m^{*2}}{2} F(0)\right)$$

$$= \left(\frac{2\pi}{m^{*2}F(0)}\right)^{3/2} \exp\left(\frac{[\mathbf{p} - m^*\mathcal{V}(t)]^2}{m^{*2}F(0)}\right).$$

Thus, if the initial statistical operator for a model system has the form $D_{t_0} = \rho(S)D_L(\Sigma)$, and, moreover, $\rho(S)$ does not depend on either V or t_0, then the corresponding distribution function of momentum $\mathbf{p}$ in the limit $V \to \infty$, i. e.

$$\lim_{V\to\infty} \omega_t(\mathbf{p}) = \lim_{V\to\infty} D_t,$$

converges to the stationary distribution function:

$$\lim_{V\to\infty}\left[\omega_t(\mathbf{p}) - \left(\frac{1}{m^*}\right)^3 \left(\frac{2\pi}{F(0)}\right)^{3/2} \exp\left(\frac{[\mathbf{p} - m^*\mathcal{V}(t)]^2}{m^{*2}F(0)}\right)\right]_{t\to\infty} = 0.$$
(3.146)

As can be seen, this stationary distribution function of momentum $\mathbf{p}$ is represented by a "shifted" Maxwellian function.

Thus the use of the Hamiltonian $\widetilde{H}^{(L)}$ as the approximate one is bound up with the assumption that a shifted Maxwellian distribution may be used for the initial approximation.

Appendix I

Lemma. For the averaged product of the operators $\tilde{b}_k(t)$ and $\mathcal{U}(S, \Sigma)$ the following equality holds

$$\mathrm{Tr}_{(S,\Sigma)} \tilde{b}_k(t)\mathcal{U}(S, \Sigma)D_{t_0}$$

$$= \frac{1}{1 - e^{-\beta\hbar\omega(k)}} \mathrm{Tr}_{(S,\Sigma)} \{\tilde{b}_k(t)\mathcal{U}(S, \Sigma) - \mathcal{U}(S, \Sigma)\tilde{b}_k(t)\}D_{t_0},$$

$$D_{t_0} = \rho(S)D_L(\Sigma).$$

Proof. Note that the Bose operators b_k commute with arbitrary operators of the electron subsystem $\Phi(S)$. Constructing an operator expression averaged with the statistical operator of the whole system, i.e. $D_{t_0} = \rho(S)D(\Sigma)$; we have

$$\mathrm{Tr}_{(S,\Sigma)} \tilde{b}_k(t)\mathcal{U}(S, \Sigma)D_{t_0} = \mathrm{Tr}_{(S,\Sigma)} \tilde{b}_k(t)\mathcal{U}(S, \Sigma)\rho(S)D(\Sigma)$$

$$= \mathrm{Tr}_{(\Sigma)} \tilde{b}_k(t)\{\mathrm{Tr}_{(S)} \mathcal{U}(S, \Sigma)\rho(S)\}D(\Sigma),$$

$$\mathrm{Tr}_{(S,\Sigma)} \mathcal{U}(S, \Sigma)\tilde{b}_k(t)D_{t_0} = \mathrm{Tr}_{(S,\Sigma)} \mathcal{U}(S, \Sigma)\tilde{b}_k(t)\rho(S)D(\Sigma)$$

$$= \mathrm{Tr}_{(\Sigma)} \{\mathrm{Tr}_{(S)} \mathcal{U}(S, \Sigma)\rho(S)\}\tilde{b}_k(t)D(\Sigma).$$

Denote

$$\mathrm{Tr}_{(S)} \mathcal{U}(S, \Sigma)\rho(S) = B(\Sigma);$$

then

$$\mathrm{Tr}_{(S,\Sigma)} \tilde{b}_k(t)\mathcal{U}(S, \Sigma)D_{t_0} = \mathrm{Tr}_{(\Sigma)} \tilde{b}_k(t)B(\Sigma)D(\Sigma),$$

$$\mathrm{Tr}_{(S,\Sigma)} \mathcal{U}(S, \Sigma)\tilde{b}_k(t)D_{t_0} = \mathrm{Tr}_{(\Sigma)} B(\Sigma)\tilde{b}_k(t)D(\Sigma). \tag{A.1}$$

It is worth recalling here an important property of the equilibrium Gibbs averages in statistical mechanics. Consider an isolated dynamical system described by some time-independent Hamiltonian H and two dynamical variables A and B relating to this system, which are also time-independent. Then, for the equilibrium averages

$$\langle A(t)B\rangle_{\mathrm{eq}} = \mathrm{Tr}\, A(t)BD_{\mathrm{eq}}, \quad \langle BA(t)\rangle_{\mathrm{eq}} = \mathrm{Tr}\, BA(t)D_{\mathrm{eq}}$$

(D_{eq} — heat bath equilibrium statistical operator), in which

$$A(t) = e^{\frac{i}{\hbar}Ht}A(0)e^{-\frac{i}{\hbar}Ht},$$

163

we have

$$\langle A(t)B\rangle_{\text{eq}} = \int_{-\infty}^{+\infty} J(\omega)e^{-i\omega t}\, d\omega, \quad \langle BA(t)\rangle_{\text{eq}} = \int_{-\infty}^{+\infty} e^{-\beta\hbar\omega} J(\omega)e^{-i\omega t}\, d\omega.$$

We write these relations in the form

$$\text{Tr}\left(e^{\frac{i}{\hbar} H(t-t_0)} A e^{-\frac{i}{\hbar} H(t-t_0)} B D_{\text{eq}}\right) = \int_{-\infty}^{+\infty} J(\omega)e^{-i\omega(t-t_0)}\, d\omega,$$

$$\text{Tr}\left(B e^{\frac{i}{\hbar} H(t-t_0)} A e^{-\frac{i}{\hbar} H(t-t_0)} D_{\text{eq}}\right) = \int_{-\infty}^{+\infty} e^{-\beta\hbar\omega} J(\omega)e^{-i\omega(t-t_0)}\, d\omega. \tag{A.2}$$

Let us put now

$$H = H(\Sigma), \quad D_{\text{eq}} = D(\Sigma), \quad A = b_k, \quad B = B(\Sigma).$$

In this case,

$$\widetilde{b}_k(t) = e^{-i\omega(k)(t-t_0)} b_k = e^{\frac{i}{\hbar} H(t-t_0)} b_k e^{-\frac{i}{\hbar} H(t-t_0)}.$$

Hence (A.2) can be transformed into

$$\text{Tr}_{(\Sigma)}\, \widetilde{b}_k(t)B(\Sigma)D(\Sigma) = e^{-i\omega(k)(t-t_0)}\, \text{Tr}_{(\Sigma)}\, \widetilde{b}_k B(\Sigma)D(\Sigma)$$

$$= \int_{-\infty}^{+\infty} J_k(\omega)e^{-i\omega(t-t_0)}\, d\omega,$$

$$\text{Tr}_{(\Sigma)}\, \widetilde{B}(\Sigma)\widetilde{b}_k(t)D(\Sigma) = e^{-i\omega(k)(t-t_0)}\, \text{Tr}_{(\Sigma)}\, \widetilde{B}(\Sigma)b_k D(\Sigma)$$

$$= \int_{-\infty}^{+\infty} e^{-\beta\hbar\omega} J_k(\omega)e^{-i\omega(t-t_0)}\, d\omega. \tag{A.3}$$

These relations show that $J_k(\omega)$ is proportional to $\delta(\omega - \omega(k))$:

$$J_k(\omega) = I_k\delta(\omega - \omega(k)),$$

and from here

$$e^{-\beta\hbar\omega} J_k(\omega) = e^{-\beta\hbar\omega(k)} J_k(\omega).$$

Hence we get from (A.3)

$$\text{Tr}_{(\Sigma)}\, B(\Sigma)\widetilde{b}_k(t)D(\Sigma) = e^{-\beta\hbar\omega(k)}\, \text{Tr}_{(\Sigma)}\, \widetilde{b}_k(t)B(\Sigma)D(\Sigma).$$

Using (A.1),

$$\text{Tr}_{(S,\Sigma)}\, \mathcal{U}(S,\Sigma)\widetilde{b}_k(t)D_{t_0} = e^{-\beta\hbar\omega(k)}\text{Tr}_{(S,\Sigma)}\, \widetilde{b}_k(t)\mathcal{U}(S,\Sigma)D_{t_0},$$

which gives

$$\text{Tr}_{(S,\Sigma)}\{\widetilde{b}_k(t)\mathcal{U}(S,\Sigma) - \mathcal{U}(S,\Sigma)\widetilde{b}_k(t)\}D_{t_0}$$

$$= (1 - e^{-\beta\hbar\omega(k)})\text{Tr}_{(S,\Sigma)}\, \widetilde{b}_k(t)\mathcal{U}(S,\Sigma)D_{t_0}.$$

We observe now that

$$\text{Tr}_{(S,\Sigma)}\widetilde{b}_k(t)\mathcal{U}(S,\Sigma)D_{t_0}$$

$$= \frac{1}{1 - e^{-\beta\hbar\omega(k)}}\,\text{Tr}_{(S,\Sigma)}\{\widetilde{b}_k(t)\mathcal{U}(S,\Sigma) - \mathcal{U}(S,\Sigma)\widetilde{b}_k(t)\}D_{t_0},$$

$$\text{Tr}_{(S,\Sigma)}\mathcal{U}(S,\Sigma)\widetilde{b}_k(t)D_{t_0} \tag{A.4}$$

$$= \frac{e^{-\beta\hbar\omega(k)}}{1 - e^{-\beta\hbar\omega(k)}}\,\text{Tr}_{(S,\Sigma)}\{\widetilde{b}_k(t)\mathcal{U}(S,\Sigma) - \mathcal{U}(S,\Sigma)\widetilde{b}_k(t)\}D_{t_0}.$$

Thus the lemma is proved.

Let us introduce the notation

$$\frac{e^{-\beta\hbar\omega(k)}}{1 - e^{-\beta\hbar\omega(k)}} = N_k.$$

Then the relations (A.4) can be expressed in terms of the phonon occupation numbers $b_k^\dagger b_k$:

$$\text{Tr}_{(S,\Sigma)}\widetilde{b}_k(t)\mathcal{U}(S,\Sigma)D_{t_0} = (1 + N_k)\text{Tr}_{(S,\Sigma)}\{\widetilde{b}_k(t)\mathcal{U}(S,\Sigma) - \mathcal{U}(S,\Sigma)\widetilde{b}_k(t)\}D_{t_0},$$

$$\text{Tr}_{(S,\Sigma)}\mathcal{U}(S,\Sigma)\widetilde{b}_k(t)D_{t_0} = N_k\text{Tr}_{(S,\Sigma)}\{\widetilde{b}_k(t)\mathcal{U}(S,\Sigma) - \mathcal{U}(S,\Sigma)\widetilde{b}_k(t)\}D_{t_0}.$$

Observation. From the proof outlined above, the suspicion might have arisen that the operator $\mathcal{U}(S,\Sigma)$ should not depend explicitly on the time t. However, it is not difficult to see that the validity of the lemma in the general case of an explicitly time-dependent operator follows immediately from its correctness for the time-independent case.

Indeed, consider an operator $\mathcal{U}(t,S,\Sigma)$ and fix $t = t_1$. In this case, $\mathcal{U}(t,S,\Sigma)$ does not depend on time from a formal point of view, and the relations (A.4) hold. Because the time t_1 can be fixed arbitrarily, one can put $t_1 = t$ in (A.4) and satisfy oneself as to their correctness.

Appendix II

It will be noted that our statement regarding (3.123) and (3.124) will be proved if we are able to prove the following lemma.

Lemma. Let $\mathbf{A}_N$ and $\mathbf{B}_N$ be a sequence of real three-dimensional vectors converging to finite limits as $N \to \infty$:

$$\mathbf{A}_N \to \mathbf{A}, \quad \mathbf{B}_N \to \mathbf{B} \quad \text{as} \quad N \to \infty. \qquad (B.1)$$

So, if our statistical operator $\rho(S)$ does not depend on N, then

$$\mathrm{Tr}_{(S)}\, e^{i(\mathbf{A}_N \cdot \mathbf{r} + \mathbf{B}_N \cdot \mathbf{p})} \rho(S) \to \mathrm{Tr}_{(S)}\, e^{i(\mathbf{A} \cdot \mathbf{r} + \mathbf{B} \cdot \mathbf{p})} \rho(S) \qquad (B.2)$$

In the case of (3.123), we have to put in this lemma, $N = V$,

$$\mathbf{B}_N = -\mathbf{k} \int_\tau^t g_0(\delta, \sigma - t_0)\, d\sigma, \qquad \mathbf{A}_N = \mathbf{k} \int_\tau^t g_1(\delta, \sigma - t_0)\, d\sigma,$$

$$\mathbf{B} = -\mathbf{k} \int_\tau^t \psi_0(\sigma - t_0)\, d\sigma, \qquad \mathbf{A} = \mathbf{k} \int_\tau^t \psi_1(\sigma - t_0)\, d\sigma.$$

And in the case of (3.124), we can put $N = t - t_0$, with τ fixed, and

$$\mathbf{B}_N = -\mathbf{k} \int_\tau^t \psi_0(\sigma - t_0)\, d\sigma, \qquad \mathbf{A}_N = \mathbf{k} \int_\tau^t \psi_1(\sigma - t_0)\, d\sigma,$$

$$\mathbf{B} = 0, \qquad \mathbf{A} = 0.$$

Let us turn to the proof of the lemma.

Proof. As long as the commutators of the components of vector operators $\mathbf{p}$ and $\mathbf{r}$ are c-numbers,

$$p_\alpha r_\beta - r_\beta p_\alpha = -i\hbar \delta_{\alpha\beta},$$

then, according to the well-known identity

$$e^{i(\mathbf{C}_1 \cdot \mathbf{r} + \mathbf{C}_2 \cdot \mathbf{p})} = e^{\frac{i\hbar}{2} \mathbf{C}_1 \cdot \mathbf{C}_2} e^{i\mathbf{C}_1 \cdot \mathbf{r}} e^{i\mathbf{C}_2 \cdot \mathbf{p}}$$

we can write:

$$e^{i(\mathbf{A}_N \cdot \mathbf{r} + \mathbf{B}_N \cdot \mathbf{p})} - e^{i(\mathbf{A} \cdot \mathbf{r} + \mathbf{B} \cdot \mathbf{p})} = e^{\frac{i\hbar}{2} \mathbf{A}_N \cdot \mathbf{B}_N} e^{i\mathbf{A}_N \cdot \mathbf{r}} e^{i\mathbf{B}_N \cdot \mathbf{p}}$$

$$- e^{\frac{i\hbar}{2} \mathbf{A} \cdot \mathbf{B}} e^{i\mathbf{A} \cdot \mathbf{r}} e^{i\mathbf{B} \cdot \mathbf{p}} = \left(e^{\frac{i\hbar}{2} \mathbf{A}_N \cdot \mathbf{B}_N} - e^{\frac{i\hbar}{2} \mathbf{A} \cdot \mathbf{B}} \right) e^{i\mathbf{A}_N \cdot \mathbf{r}} e^{i\mathbf{B}_N \cdot \mathbf{p}}$$

$$+ e^{\frac{i\hbar}{2} \mathbf{A} \cdot \mathbf{B}} \left[e^{i\mathbf{A}_N \cdot \mathbf{r}} \left(e^{i\mathbf{B}_N \cdot \mathbf{p}} - e^{i\mathbf{B} \cdot \mathbf{p}} \right) + \left(e^{i\mathbf{A}_N \cdot \mathbf{r}} - e^{i\mathbf{A} \cdot \mathbf{r}} \right) e^{i\mathbf{B} \cdot \mathbf{p}} \right].$$

But, because the operator $e^{i\mathbf{A}_N \cdot \mathbf{r}} e^{i\mathbf{B}_N \cdot \mathbf{p}}$ is unitary owing to the reality of $\mathbf{A}_N, \mathbf{B}_N$, we have

$$\left| \mathrm{Tr}_{(S)} \, e^{i\mathbf{A}_N \cdot \mathbf{r}} e^{i\mathbf{B}_N \cdot \mathbf{p}} \rho(S) \right| \leqslant 1,$$

and hence

$$\left| \mathrm{Tr}_{(S)} \left(e^{i(\mathbf{A}_N \cdot \mathbf{r} + \mathbf{B}_N \cdot \mathbf{p})} \rho(S) \right) - \mathrm{Tr}_{(S)} \left(e^{i(\mathbf{A} \cdot \mathbf{r} + \mathbf{B} \cdot \mathbf{p})} \rho(S) \right) \right|$$

$$\leqslant \left| e^{\frac{i\hbar}{2}(\mathbf{A}_N \cdot \mathbf{B}_N - \mathbf{A} \cdot \mathbf{B})} - 1 \right| + \left| \mathrm{Tr}_{(S)} \, e^{i\mathbf{A}_N \cdot \mathbf{r}} \left(e^{i\mathbf{B}_N \cdot \mathbf{p}} - e^{i\mathbf{B} \cdot \mathbf{p}} \right) \rho(S) \right|$$

$$+ \left| \mathrm{Tr}_{(S)} \left(e^{i\mathbf{A}_N \cdot \mathbf{r}} - e^{i\mathbf{A} \cdot \cdot \mathbf{r}} \right) e^{i\mathbf{B} \cdot \mathbf{p}} \rho(S) \right|. \quad (B.3)$$

Bearing in mind that, since it is statistical operator, $\rho(S)$ is non-negative, the following general inequality holds (see Appendix III):

$$\left| \mathrm{Tr}_{(S)} \, UV\rho(S) \right|^2 \leqslant \mathrm{Tr}_{(S)} \, UU^\dagger \rho(S) \, \mathrm{Tr}_{(S)} \, V^\dagger V \rho(S). \quad (B.4)$$

With the help of this inequality, we evaluate the first and second expressions with the symbol $\mathrm{Tr}_{(S)}$ on the right-hand side of (B.3), choosing

$$U_I = e^{i\mathbf{A}_N \cdot \mathbf{r}}, \; V_I = \left(e^{i\mathbf{B}_N \cdot \mathbf{p}} - e^{i\mathbf{B} \cdot \mathbf{p}} \right), \; U_{II} = \left(e^{i\mathbf{A}_N \cdot \mathbf{r}} - e^{i\mathbf{A} \cdot \mathbf{r}} \right), \; V_{II} = e^{i\mathbf{B} \cdot \mathbf{p}}.$$

It can be seen that V_{II} and U_I are unitary, and because

$$\mathrm{Tr}_{(S)} \, \rho(S) = 1,$$

for any statistical operator, we have

$$\mathrm{Tr}_{(S)} \, U_I U_I^\dagger \rho(S) = \mathrm{Tr}_{(S)} \, V_{II}^\dagger V_{II} = 1.$$

Furthermore, because the components of the vector $\mathbf{r}$ commute with each other, as do the components of the vector $\mathbf{p}$, we get

$$U_{II} U_{II}^\dagger = 2[1 - \cos(\mathbf{A}_N - \mathbf{A}) \cdot \mathbf{r}],$$

$$V_I^\dagger V_I = 2[1 - \cos(\mathbf{B}_N - \mathbf{B}) \cdot \mathbf{p}].$$

Therefore we find from (B.3) that

$$\left| \mathrm{Tr}_{(S)} \left(e^{i\mathbf{A}_N \cdot \mathbf{r} + \mathbf{B}_N \cdot \mathbf{p}} \rho(S) \right) - \mathrm{Tr}_{(S)} \left(e^{i\mathbf{A} \cdot \mathbf{r} + \mathbf{B} \cdot \mathbf{p}} \rho(S) \right) \right|$$

$$\leqslant \left[2 \left(1 - \cos \frac{\hbar}{2} (\mathbf{A}_N \cdot \mathbf{B}_N - \mathbf{A} \cdot \mathbf{B}) \right) \right]^{1/2}$$

$$+ \left\{ \mathrm{Tr}_{(S)} \, 2[1 - \cos(\mathbf{B}_N - \mathbf{B}) \cdot \mathbf{p}] \rho(S) \right\}^{1/2}$$

$$+ \left\{ \mathrm{Tr}_{(S)} \, 2[1 - \cos(\mathbf{A}_N - \mathbf{A}) \cdot \mathbf{r}] \rho(S) \right\}^{1/2}. \quad (B.5)$$

Consider matrix elements of $\rho(S)$ in the r-representation, i.e. $\langle \mathbf{r}|\rho(S)|\mathbf{r}\rangle$, and in the p-representation, $\langle \mathbf{p}|\rho(S)|\mathbf{p}\rangle$. Then

$$\mathrm{Tr}_{(S)}(1 - \cos{(\mathbf{B}_N - \mathbf{B})} \cdot \mathbf{p})\rho(S) = \int[1 - \cos{(\mathbf{B}_N - \mathbf{B})} \cdot \mathbf{p}]\langle \mathbf{p}|\rho(S)|\mathbf{p}\rangle \, d\mathbf{p},$$

$$\mathrm{Tr}_{(S)}[1 - \cos{(\mathbf{A}_N - \mathbf{A})} \cdot \mathbf{p}]\rho(S) \int[1 - \cos{(\mathbf{A}_N - \mathbf{A})} \cdot \mathbf{r}]\rho(S)\langle \mathbf{r}|\rho(S)|\mathbf{r}\rangle \, d\mathbf{r}.$$

But diagonal elements of a non-negative operator are non-negative:

$$\langle \mathbf{p}|\rho(S)|\mathbf{p}\rangle \geqslant 0, \quad \langle \mathbf{r}|\rho(S)|\mathbf{r}\rangle \geqslant 0,$$

and, because of the identity $\mathrm{Tr}_{(S)}\,\rho(S) = 1$, we have

$$\int\langle \mathbf{p}|\rho(S)|\mathbf{p}\rangle = 1, \quad \int\langle \mathbf{r}|\rho(S)|\mathbf{r}\rangle = 1.$$

Taking into account that $\rho(S)$ does not depend on N and also that

$$1 - \cos X \leqslant 2, \ 1 - \cos{(\mathbf{A}_N - \mathbf{A})} \cdot \mathbf{r} \to 0 \ \text{при } N \to \infty$$

for bounded $\mathbf{r}$ and

$$1 - \cos{(\mathbf{B}_N - \mathbf{B})} \cdot \mathbf{p} \to \infty \ \text{as } N \to \infty$$

for bounded $\mathbf{p}$, and

$$\mathrm{Tr}_{(S)}[1 - \cos{(\mathbf{A}_N - \mathbf{A})} \cdot \mathbf{p}]\rho(S) \to 0 \ \text{as } N \to \infty,$$

$$\mathrm{Tr}_{(S)}[1 - \cos{(\mathbf{B}_N - \mathbf{B})} \cdot \mathbf{p}]\rho(S) \to 0 \ \text{as } N \to \infty.$$

Hence we conclude on the basis of (B.5) that the lemma is proved. It is obvious that the inequality (B.5) is valid whether or not $\rho(S)$ depends on N. Besides that, it is clear that $2(1 - \cos x) < x^2$. Hence

$$\left| \mathrm{Tr}_{(S)}\left\{ e^{i(\mathbf{A}_N \cdot \mathbf{r} + \mathbf{B}_N \cdot \mathbf{p})}\rho(S) \right\} - \mathrm{Tr}_{(S)}\left\{ e^{i(\mathbf{A} \cdot \mathbf{r} + \mathbf{B} \cdot \mathbf{p})}\rho(S) \right\} \right|$$

$$\leqslant \frac{\hbar}{2}|\mathbf{A}_N \cdot \mathbf{B}_N - \mathbf{A} \cdot \mathbf{B}| + [\mathrm{Tr}_{(S)}\,|\mathbf{p}|^2\rho(S)]^{1/2}|\mathbf{B}_N - \mathbf{B}|$$

$$+ |\mathbf{A}_N - \mathbf{A}|[\mathrm{Tr}_{(S)}\,|\mathbf{r}|^2\rho(S)]^{1/2}.$$

Thus, if $\rho(S)$ depends on N, such that

$$\mathrm{Tr}_{(S)}\,|\mathbf{p}|^2\rho(S) \leqslant K_1^2, \quad \mathrm{Tr}_{(S)}\,|\mathbf{r}|^2\rho(S) \leqslant K_2^2,$$

where K_1 and K_2 do not depend on N, then (B.2) holds. Therefore, if in Chapter 3 $\rho(S)$ depends on t_0 and V, but in such a way that

$$\langle \mathbf{p}^2 \rangle_{t_0} = \mathrm{Tr}_{(S)}\,|\mathbf{p}|^2\rho(S), \quad \langle \mathbf{r}^2 \rangle_{t_0} = \mathrm{Tr}_{(S)}\,|\mathbf{r}|^2\rho(S)$$

are bounded by some magnitudes independent of V or t_0, then all the speculations and conclusions of Chapter 3 remain valid.

Appendix III

Let us consider an average of a two-operator product $\langle AB \rangle$ as a bilinear form in A and B (linear with respect to each of these operators).

Let $Z(A, B)$ be an arbitrary bilinear form in A and B with the following properties:

$$Z(A^\dagger, A) \geqslant 0, \tag{C.1}$$

$$\{Z(A, B)\}^* = Z(B^\dagger, A^\dagger). \tag{C.2}$$

We are going to show that the following inequality always holds:

$$|Z(A, B)|^2 \leqslant Z(A, A^\dagger) Z(B^\dagger, B). \tag{C.3}$$

Putting here

$$A = U[\rho(S)]^{1/2},$$

$$B = V[\rho(S)]^{1/2},$$

we arrive at the inequality (B.4).

To prove this inequality, let us not, first of all, that, thanks to (C.1),

$$Z(xA + y^*B^\dagger, x^*A^\dagger + yB) \geqslant 0, \tag{C.4}$$

where x and y are arbitrary numbers. Removing the parentheses, we get

$$xx^* Z(A, A^\dagger) + xy Z(A, B) + y^*x^* Z(B^\dagger, A^\dagger) + y^*y Z(B^\dagger, B) \geqslant 0.$$

Choose for x, y, x^*, y^*

$$x^* = -Z(A, B),$$

$$x = -\{Z(A, B)\}^* = -Z(B^\dagger, A^\dagger),$$

$$y = y^* = Z(A, A^\dagger).$$

Then

$$-|Z(A, B)|^2 Z(A, A^\dagger) + [Z(A, A^\dagger)]^2 Z(B^\dagger, B) \geqslant 0.$$

From here, if $Z(A, A^\dagger) \neq 0$, we get the inequality (C.3).

It only remains to show that if $Z(A, A^\dagger) = 0$, then

$$Z(A, B) = 0. \tag{C.5}$$

171

For this purpose, we put in (C.4)

$$x^* = -Z(A, B)R,$$
$$x = -Z(B^\dagger, A^\dagger)R,$$
$$y = y^* = 1,$$

where R is an arbitrary positive number. We find that

$$-2R|Z(A, B)|^2 + Z(B^\dagger, B) \geqslant 0. \qquad (C.6)$$

Let $R \to \infty$. Then, if (C.5) is wrong, we see that the left-hand side of (C.6) must approach $-\infty$, which is impossible.

References

1. Pekar S.I. //Sov. Phys. JETP, 1946. V. 16. P. 341; see also Pekar S.I. Issledovania po electronnoi teorii kristallov. — M.: Gostekhizdat, 1951. (in Russian)

2. Landau L.D. // Phys. Z. Sowietunion. 1933. V. 3. P. 664.

3. Landau L.D., Pekar S.I. // Sov. Phys. JETP, 1948. V. 18. P. 419.

4. Fröhlich H., Pelzer H., Zienau S. // Philos. Mag. 1950. V. 41. P. 221.

5. Feynman R.P. Slow electrons in a polar crystal // Phys. Rev. 1955. V. 97. P. 660–665; see. also Feynman R.P. Statistical Mechanics. — Benjamin, Reading, MA, 1972.

6. In: Polarons in Ionic Crystals and Polar Semiconductors / Ed. J.T. Devreese. — North–Holland. — Amsterdam, 1972; see also Devreese J.T., Evrard R. Linear and Nonlinear Transport in Solids. — N.Y.: Plenum Press, 1976.

7. Bogolubov N.N. New adiabatic form of perturbation theory in the problem of interaction of a particle with quantum field // Ukrainian Math. J. 1950. V. II. No. 2.

8. Bogolubov N.N., Bogolubov N.N. Jr. Aspects of the polaron theory. JINR Communications, P-17-81-85, Dubna, 1981; Some Aspects of Polaron Theory. V. 4. — World Scientific, 1988; Foundations of Physics. 1985. V. 15. № 11; Plenum Press. FNDPA4. 1985. V. 15(11). ISSN 0015-9018.

9. Bogolubov N.N. Jr., Kireev A.N. Ground-state energy of the surface polaron. In: Proceedings of the 1st Jagna International Workshop on Advances in Theoretical Physics (ed. Bernido C., Carpio–Bernido M.V.). — Jagna, Philippines, 1996. P. 90–109.

10. Bogolubov N.N. Jr., Kireev A.N. Surface polaron with a barrier potential: upper bound to the ground-state energy // Int. J. Mod. Phys. 1996. V. B. 10. P. 455–470.

11. Appel J. // Solid State Phys. 1968. V. 21. P. 193.

12. Iadonisi G. Electron–phonon interaction: effects on the excitation spectrum of solids // Riv. Nuovo Cimento. 1984. V. 7, № 11.

13. Lee T.D., Low F.E., Pines D. The motion of slow electrons in a polar crystal // Phys. Rev. 1953. V. 90. P. 297.

14. In: Proceedings of the XV International Conference on Physics of Semiconductors (Kyoto, 1980) // J. Phys. Soc. Jpn., Suppl. 1985. V. A49.

15. In: Recent Developments in Condensed Matter Physics. V. 1 / Ed. J.T. Devreese. — N.Y.: Plenum, 1981.

16. In: Physics of Polarons and Excitons in Polar Semiconductors and Ionic Crystals / Eds. J.T. Devreese, F. Peeters. — N.Y.: Plenum, 1981.

17. Rodriguez K., Fedyanin V.K. Method of continual integration in the polaron problem // Physics of Elementary Particles and Atomic Nuclei (PEPAN), (JINR, Dubna), 1984. V. 15. P. 870.

18. Gerlach B., Lowen H. Analytical properties of polaron system // Rev. Mod. Phys. 1991. V. 63. P. 63–90.

19. Tyablikov S.V. To a theory of interaction of a particle with a quantum field // Sov. Phys. JETP, 1951. V. 21. P. 16.

20. In: Polarons and Excitons, Scottish Universities' Summer School in Physics, St. Andrews (1962) / Eds. C.G. Kuper, G.D. Whitfield, Edinburgh, London: Oliver and Boyd, 1963.

21. Tulub A.V. Slow electrons in polar crystals // Sov. Phys. JETP, 1961. V. 41. P. 1828.

22. Hohler G., Mullensiefen A. Störungstheoretische Berechnung der Selbsenergie und der Masse des Polaron // Z. Phys. 1959 S. 157. P. 159.

23. Roseler J. A new variational ansatz in the polaron theory // Phys. Status Solidi. 1968. V. 25. P. 311.

24. Larsen D.M. Upper and lower bounds for the intermediate-coupling polaron ground-state energy // Phys. Rev. V. 172. P. 967 (1968).

25. Matz D., Burkey B.C. Dynamical theory of the large polaron: Fock approximation // Phys. Rev. 1971. V. B3. P. 3487.

26. Miyake S. Strong coupling limit of the polaron ground state // J. Phys. Soc. Jpn. 1975. V. 38. P. 181–182.

27. Krivoglaz M.A., Pekar S.I. // Fortschr. Phys. 1961. V. 4. P. 73; see also Krivoglaz M.A., Pekar S.I. // Bull. Acad. Sci. USSR, Ser. Phys. 1957. V. 21. P. 3.

28. Osaka Y. Polaron state at finite temperature // Progr. Theor. Phys. 1959. V. 22. P. 437.

29. Feynman R.P., Hellwarth R.W., Iddings C.K., Platzman P.M. Mobility of slow electrons in a polar crystal // Phys. Rev. 1962. V. 127. P. 1004.

30. Thornber K.K., Feynman R.P. Velocity acquired by an electron in a finite electric field in a polar crystal // Phys. Rev. 1970. V. B1. P. 4099.

31. Thornber K.K. Linear and nonlinear electronic transport in electron–phonon systems: self-consistent approach within the path-integral formalism // Phys. Rev. 1971. V. B3. P. 1929.

32. Klyukanov A.A., Pokatilov E.P. Electric conductivity tensor for polarons in a magnetic field // Sov. Phys. JETP, 1971. V. 60. P. 312.

33. Bogolubov N.N. Kinetic equations for the electron–phonon system. JINR Communications, E17-11822, Dubna, 1978.

34. Bogolubov N.N. Jr. Kinetic equation for a dynamical system interacting with a phonon field // Sov. Phys.Theor. Math. Phys. 1979. V. 40. P. 77.

35. Bogolubov N.N.,Bogolubov N.N. Jr. Kinetic equation for dynamical systems interacting with phonon field // Physics of Elementary Particles and Atomic Nuclei (PEPAN), (JINR, Dubna) 1981. V. 11, No. 2. P. 245–300.

36. Bogolubov N.N. Jr., Plechko V.N. Perturbation theory for polaron model at finite temperature. // Sov. Phys. Theor. Math. Phys. 1985. V. 65. P. 423–434; Approximation Methods in the Polaron Theory // Rivisto del Nuovo Cimento. 1988. V. 11. № 9.

37. Proceedings of the 1979 International Summer School on New Developments in Semiconductor Physics, held in Szeged, Hungary. — West Berlin: Springer Verlag, 1980.

38. Path Integrals and their Applications in Quantum, Statistical and Solid State Physics / Eds. G.J. Papadopoulos, J.T. Devreese. — N.Y.: Plenum Press, 1978.

39. Feynman R.P., Hibbs A.R. Quantum Mechanics and Path Integrals. Copyright 1965, Library of Congress Catalog Card Number, 64–25171.

40. Bogolubov N.N. Jr., Sadovnikov B.I. // Sov. Phys. JETP, 1962. V. 43, No. 8.

41. Bogolubov N.N. Jr., Sadovnikov B.I. Some Problems in Statistical Mechanics. — M.: Vysshaya Shkola, 1975.

42. Solodovnikova E.P., Tavkhelidze A.N., Khrustalev O.A. Oscillator levels of a particle as a consequence of strong interaction with a field // Sov. Phys., Theor. Math. Phys.1972. V. 10. P. 162–181.

43. Solodovnikova E.P., Tavkhelidze A.N., Khrustalev O.A., Bogolubov's transformation in the theory of strong coupling. II // Sov. Phys., Theor. Math. Phys. 1972. V. 11. P. 317–330.

44. Fedyanin V.K., Mochinsky B.V., Rodriguez C. Generating Functional and Functional Analog of the Variational Principle of N.N Bogolubov. E17-12850. — Dubna: JINR, 1979.

45. Kochetov E.A., Kuleshov S.P., Smondyrev M.A. Functional integration

46. Vo Hong Ahn. A. Quantum Approach to the Parametric Excitation Problem in Solids // Physics Reports. 1980. V. 64, No. 1. P. 1–45.

47. Bogolubov N.N. Selected Works. V. 3. — Kiev: Naukova Dumka, 1971. P. 213.

48. Ziman J.M. Electrons and Phonons. Clarendon Press. — Oxford, 1960.

49. Krylov N.M., Bogolubov N.N. Proceedings of the chair of mathematical physics V. 4. — Kiev: Academy of Sciences of Ukraine, 1939. P. 5.

50. Bogolubov N.N. On Some Statistical Methods in Mathematical Physics. — Kiev, Publishing House of the Academy of Sciences of Ukraine, 1945. see also Bogoliubov N.N. in: Studies in Statistical Mechanics. V. 3 / Ed. J. de Boer and G.R. Uhlenbeck. — Amsterdam: North-Holland Publ. Co., 1962.

51. Shelest A.V. Preprint ITP. 1967. P. 67–11; N.N. Bogolubov's Method in the Dynamic Theory of Kinetic Equations. — M.: Nauka, 1990.

52. Bogolubov N.N. On Stochastic Processes in Dynamic Systems // PEPAN (JINR, Dubna). 1978. V. 9, No. 4.

53. Bogolubov N.N. Jr., Fam Le Kien, Shumovsky A.S. // Sov. Phys., Theor. Math. Phys. 1982. V. 52. P. 423; Sov. Phys., Theor. Math. Phys. 1984. V. 60. P. 254.

54. Bogolubov N.N. Jr., Kazaryan A.R., Kurbatov A.M., Neskoromny V.N. // Sov. Phys., Theor. Math. Phys. 1983. V. 54. P. 147; Sov. Phys., Theor. Math. Phys. 1984. V. 59. P. 249.

55. Kazaryan A.R. // Bull. Acad. Sci. USSR, Ser. Phys. 1981. V. 258. P. 336.

56. Bogolubov N.N. Jr., Sadovnikov B.I., Shumovsky A.S. Mathematical Methods of Statistical Mechanics of Model Systems. — M.: Nauka, 1989.

57. Peeters F.M., Devreese J.T. // Sol. St. Comm. 1981. V. 39, No. 3. P. 445-449.

58. Peeters F.M., Devreese J.T. // Phys. Rev. B. 1982. V. 25, No. 12. P. 7281-7301. P. 7302-7326.

59. Chernikov S.N., Rodriguez C., Fedyanin V.K. // N.N. Bogolubov's linear model for polaron in a magnetic field. P17-83-95. - Dubna: JINR, 1983.

60. Fedyanin V.K., Rodrigues C. // Physica A. 1982. V. 112, No. 3. P. 615-630.

61. Shimoda Y., Nakajima T., Sawada A. // Mod. Phys. Lett. B. 2005. V. 19, No. 11. P. 539-548.

About the Authors

Academician N. N. Bogolubov was among the world's most distinguished theoretical physicists and mathematicians. N. N. Bogolubov started his scientific career in Kiev (Ukraine), when he joined a seminar under academician N. M. Krylov. Having published his first scientific paper at the age 14, N.N. Bogolubov enjoyed an active career for more than six decades. In 1930, N. N. Bogolubov was awarded a prize from the Academy of Science of Bologna. In the same year, the Presidium of the USSR Academy of Science conferred on him the degree of Doctor of Mathematics. In 1948, he was elected a member of the Ukranian Academy of Sciences and in 1953, a member of the USSR Academy of Sciences. Subsequently, he became a member of the Presidium of the Academy and head of the Mathematics section. In 1964, he became Director of Joint Institute for Nuclear Research — an international center for fundamental research located at Dubna (near Moscow). He was also a member of the Russian Government as Deputat of Supreme Soviet.

N. N. Bogolubov was involved not only in important scientific research, which he continued without interruption, but also in the organization of science and teaching. He lectured at the Kiev and Moscow universities and directed many seminars. He was also an extraordinary tutor and creator of several well-known schools of mathematical physics and nonlinear mechanics in Kiev and theoretical physics in Moscow and Dubna. It is possible to mention some well-known books such as *Introduction to the Theory of Quantized Fields* (by N. N. Bogolubov and D. V. Shirkov), *An Introduction of Quantum Statistical Mechanics* (by N. N. Bogolubov and N. N. Bogolubov Jr.), *Introduction to Axiomatic Quantum Field Theory* (by N. N. Bogolubov, A. A. Logunov and I. I. Todorov) and *Problems of a Dynamical Theory in Statistical Physics* (by N. N. Bogolubov).

During his many years of scientific research in mathematics and theoretical physics, Bogolubov has written about 300 papers on nonlinear mechanics, quantum field theory, elementary particle physics, statistical physics, etc. N. N. Bogolubov also received many governmental awards (e.g. the Stalin prize, Lenin prize, Planck's golden medal in Germany, Benjamin Franklin's medal in USA, etc.)

His devotion to physics is shared by his son, N. N. Bogolubov Jr. who is currently the head of the statistical mechanics project in Steklov Mathematical Institute, Academy of Science, Russia. Professor N. N. Bogolubov Jr. was born in Kiev in 1940. His education started in Kiev and was furthered in Moscow. He attended Moscow State University from

1957–1963, became Assistant Professor from 1966–1968, Doctor of Science in 1970, full Professor in Academy of Science, Russia in 1973, Steklov Mathematical Institute in 1984, and was the corresponding member of the Russian Academy of Science. In 1983, he got the governmental prize for his work in statistical mechanics. He has worked in the International Organization UNESCO as an expert in theoretical physics, Delhi, India, 1969, Stony Brook University, NY, USA, 1972, Rockefeller University, NY, USA, as a guest Professor with E. G. D. Cohen and Mark Kac for several months. Also in the Research Institute for Fundamental Physics, Kyoto University, Japan, 1974, as a visiting Professor at Yokogama University and Belgium Antwerpen University etc.